Polyethylene Glycol as an Embedment for Microscopy and Histochemistry

Polyethylene Glycol as an Embedment for Microscopy and Histochemistry

Edited by

Kuixiong Gao, Ph.D.
Department of Anatomy and Cell Biology
University of Cincinnati College of Medicine
Cincinnati, Ohio

CRC Press
Boca Raton Ann Arbor London Tokyo

Library of Congress Cataloging-in-Publication Data

Polyethylene glycol as an embedment for microscopy and
histochemistry / edited by Kuixiong Gao.
p. cm.
Includes bibliographical references and index.
ISBN 0-8493-4323-2
1. Polyethylene glycol embedding (Microscopy) I. Gao, Kuixiong.
QH234.P64 1993
578'.6--dc20 92-41430
CIP

Direct all inquiries to CRC Press, Inc., 2000 Corporate Blvd., N.W., Boca Raton, Florida 33431.

International Standard Book Number 0-8493-4323-2

Library of Congress Card Number 92-41430

Printed in the United States of America 1 2 3 4 5 6 7 8 9 0

Printed on acid-free paper

PREFACE

The aim of scientific research is to explore the unknown, to find truth, and, in the final analysis, to benefit mankind. Proper methods are of critical importance for obtaining successful results. There is nothing in the world that is ultimately perfect or could not be further improved.

Paraffin embedding and sectioning, section staining, and the microscope are three essential ingredients that constitute the foundation of histology. The paraffin section method is good, reliable, and is still extensively used in biomedical research. However, the paraffin section is also imperfect. It has some shortcomings, i.e., the necessity of dehydration and clearing of the samples with harsh organic solvents, which may damage some antigenicity in tissues.

Polyethylene glycol (PEG), on the other hand, is water-soluble. It is especially suitable for use as an embedment for immunohistochemistry or immunocytochemistry. PEG embedding can be considered the second generation of paraffin embedding. The two methods are quite similar for most of the procedures employed, except that no organic solvents are necessary for PEG embedding. So, it is natural to choose PEG as a water-soluble and removable embedment for immunocytochemistry and lipid/enzyme histochemistry.

I appreciate the contributors to this book; they have spent several years of their lifetimes, through trial and error, to establish or improve the PEG embedding technique and now unreservedly devote these valuable experiences to the public. I believe that these experiences may be useful to the readers. Of course, all the methods introduced in this book are also imperfect; they ought to be modified and improved by the readers. What we have done is like lighting a match for those who have a strong will to further develop the PEG embedding technique. You yourselves are the masters who will keep the torch burning and emit bright light.

Finally, I should like to use this opportunity to express my sincere thanks to all those who volunteered their time to help me and to the contributors for preparing the manuscripts.

Kuixiong Gao
Cincinnati, Ohio

THE EDITOR

Kuixiong Gao, Ph.D., is the Research Assistant Professor of the Department of Anatomy and Cell Biology, University of Cincinnati College of Medicine at Cincinnati.

Dr. Gao graduated in 1962 from the Nanjing (Nanking) University, Nanjing, China, receiving his degrees, a B.S. in Zoology and a M.S. in Histology and Embryology. He obtained his Ph.D. in Immunology and Experimental Embryology in 1966 at the Chinese Academy of Sciences and then served as a Research Scientist, similar to a postdoctor in the U.S. During this time, the highlight of his research work was the establishment of several advanced techniques including the gas-liquid chromatographic detection of steroids and prostaglandin (at nanogram level for the national awarded research work on herb contraceptives) and the silanated water-soluble aminoplastic embedding medium for electron microscopy to retain lipids for study of membrane biogenesis. He was the core person of a team that produced and published the first pseudo-color electron micrograph through an image processing technique in Shanghai, in 1983.

Dr. Gao came to the U.S. in 1984. First he worked as a Visiting Faculty Assistant Professor, Biochemistry, at the University of Tennessee, Knoxville (1984 to 1986), where he invented the "Colloidal-Gold Labeled Solid Core Liposomes", that the university later acquired a patent for. He was then invited by the University of North Carolina at Chapel Hill to do research work related to immunocytochemical localization of proteins in the neuromuscular junction (Visiting Scientist in the Department of Anatomy and Visiting Professor in the Department of Physiology, 1986 to 1988). He then returned to the University of Tennessee, Knoxville, and served in the Animal Science Department as a Senior Research Scientist, using the immunogold labeling method to study the distribution of retinol-binding protein and growth factors (TGF-α, TGF-β1) in tissues (1988 to 1991). During that period he established a new method for transfer of PEG sections onto silanated slides with agarose blocks for immunocytochemistry.

Dr. Gao is a member of American Society for Cell Biology, New York Academy of Sciences, Microscopy Society of America, and Microscopy Society of the Ohio Valley.

Dr. Gao has published more than 30 research papers and has been the author or co-author (or co-translator) of 6 books, among which he is a contributor to the *Colloidal Gold: Principles, Methods, and Applications,* Volume 1 (edited by M. A. Hayat). His current major research interests include the immunocytochemical localization of glycogen metabolic enzymes in normal and diabetic livers and the improvement of antigenic detection, both sensitivity and resolution, with immunogold-silver stain and epipolarized microscopy.

CONTRIBUTORS

Arturo Alvarez-Buylla, Ph.D.
The Rockefeller University
New York, New York

Jonathan B. L. Bard, M.A., Ph.D.
Molecular Genetics Section
MRC Human Genetics Unit
Western General Hospital
Edinburgh, Scotland

David G. Capco, Ph.D.
Associate Professor
Molecular and Cellular Biology Program
Department of Zoology
Arizona State University
Tempe, Arizona

David F. Clayton, Ph.D.
Assistant Professor
Department of Cell and Structural Biology
University of Illinois at Urbana-Champaign
Urbana, Illinois

I. A. Efimov, Ph.D.
Professor
Institute of Cell Biophysics
Puschino-on-Oka
Moscow, Russia

Kuixiong Gao, Ph.D.
Research Assistant Professor
Department of Anatomy and Cell Biology
University of Cincinnati College of Medicine
Cincinnati, Ohio

Hisatake Kondo, M.D., Ph.D.
Professor
Department of Anatomy
Tohoku University School of Medicine
Sendai, Japan

Michael R. Leous, B.S., M.A.
Technical Writer
Boehringer Mannheim Corporation
Biochemical Products Division
Indianapolis, Indiana

Jeanette Mowery, Ph.D.
ExOxEmis, Inc.
San Antonio, Texas

Allyson S. A. Ross
Molecular Genetics Section
MRC Human Genetics Unit
Western General Hospital
Edinburgh, Scotland

TABLE OF CONTENTS

Dedicated to my friends, old and new . . .

especially those young people full of vitality

and creativity

who have the common wish to add bricks

to the pagoda of histology

Chapter 1

AN OVERVIEW OF THE POLYETHYLENE GLYCOLS

Michael R. Leous

TABLE OF CONTENTS

0-8493-4323-2/93/$0.00 + $.50

I. GENERAL INFORMATION ON THE POLYETHYLENE GLYCOLS

A. CHEMICAL NATURE

The polyethylene glycols (PEGs) are a family of synthetic, water-soluble polymers sharing a common chemical structure — primary hydroxyl groups at each end of a linear polyether chain consisting of multiple oxyethylene ($-CH_2-CH_2-O-$) subunits. The group features the general formula

$$HO-(CH_2CH_2O)_n-H$$

where n is a measure of the average number of oxyethylene subunits (sometimes referred to as the "degree of polymerization").

Chemical reactivity of the PEGs results from the primary hydroxyl groups residing at each end of the molecule. Although the influence of the polyether chain on the PEG increases with its length because of the increased number of ether linkages,[76] these oxyethylene-containing chains are largely unreactive. Therefore, 83 to 99% of the molecule is chemically inert.[63]

Table 1 lists the grade numbers and some properties of the more common PEGs. All data in Table 1 was obtained from *The Merck Index*.[73] This table gives the average number of oxyethylene residues in the polyether chain (average value of n). This number determines the physical state that each PEG assumes at room temperature. PEGs with lower molecular weights (200, 400, and 600) take the form of clear, viscous liquids, while those PEGs with higher molecular weights are available as free-flowing white powders (1500 and 4000) and/or creamy white flakes (4000 and 6000).[73]

TABLE 1
Properties of Some Common PEGs

PEG product number	200	400	600	1500	4000	6000
Average value of *n*	4	8.2–9.1	12.5–13.9	29–36	68–84	158–204
Molecular weight range	190–210	380–420	570–630	1300–1600	3000–3700	7000–9000
Viscosity at 210°F, in centistrokes	4.3	7.3	10.5	25–32	76–110	470–900
Freezing/melting range	Supercools	4–8°	20–25°	44–48°	54–58°	56–63°
Specific gravity ($d^{25/25}$)	1.127	1.128	1.128	1.210	1.212	1.210

Each PEG product is actually a mixture of PEGs, as is evident from the molecular weight ranges given in Table 1. The number a product bears (e.g., PEG 400) is an average molecular weight of the various polymers it contains. In other words, PEG 400 has an average molecular weight of 400, but it comprises numerous constituent PEGs that have molecular weights smaller and larger than 400.

In 1940, Flory derived a series of size-distribution equations for polymers formed by the sequential addition of monomers to a fixed number of outset molecules.[26] With these formulas, he determined that the percentages of each of the different constituent molecules in a polymer can be represented by a Poisson distribution. For example, PEG 400 consists of large percentages of PEG molecules weighing 370 (octa-ethylene glycol) or 415 (nona-ethylene glycol), as well as much smaller percentages of molecules weighing 238 (penta-ethylene glycol) or 591 (trideca-ethylene glycol), and the percentages of each of these components in PEG 400 are distributed according to Poisson's law.

B. HISTORY OF PEG PRODUCTION

The first recorded preparation of PEGs occurred almost 130 years ago when Lourenco and Wurtz independently prepared the first PEGs by heating ethylene glycol with water.[28] However, PEGs were restricted to a small group of products until the start of World War II, when esters of the PEGs and other ethylene oxides were produced in Germany to replace fats and oils consumed during the war.[44] About the same time, Union Carbide®* Corporation made the PEGs commercially available for the first time.[76] In the post-war era, the number of applications for the PEGs and their derivatives has increased dramatically, and the number of U.S. PEG suppliers has paralleled this growth.

The PEGs listed in Table 1 are available from numerous suppliers in the U.S. and abroad. Besides the ones listed here, a few additional, less common PEGs and specialized blends of two or more PEGs are available from some suppliers. Further information on any of the PEGs can be found in the numerous technical bulletins available from PEG suppliers such as the Dow® Chemical Company, Union Carbide® Corporation, and BASF® Corporation.

II. PROPERTIES OF THE PEGs

Some of the properties of the PEGs are listed in Table 1.[73] Characteristics shared by all of the PEGs include complete solubility in water, the ability to dissolve many substances, good stability, wide compatibility with other substances, blandness, low toxicity levels, and good lubricity.[76] Some of these qualities warrant further descriptions, and these are given in the following sections.

* Registered trademark of Union Carbide® Corporation, Danbury, CT.

Unless otherwise cited, the information in this section was obtained from a chapter by George M. Powell, III,[76] which provides an excellent, thorough review of all aspects of the PEGs.

A. VISCOSITY

As shown in Table 1, molecular-weight increases parallel viscosity increases in the PEGs.[63] Temperature also affects the viscosities of the PEGs; they become less viscous as temperatures increase. This is important for the higher PEGs, which display a wide viscosity range at different temperatures — a characteristic that has made them useful as lubricants capable of functioning over a wide range of temperatures.[63] It should also be noted that both the pure PEGs and aqueous solutions containing them are "Newtonian fluids" (i.e., their viscosities are independent of shear).[87]

B. FREEZING/MELTING RANGE

Table 1 also displays the freezing/melting ranges for the PEGs. Note that the freezing/melting ranges of the high molecular weight PEGs are broader than those of the liquid PEGs.[63] This occurs because, as Flory (1940) showed,[26] the solid PEGs contain a larger number of different polymer constituents.

C. HYGROSCOPICITY

The PEGs vary in their degree of hygroscopicity — the ability to absorb and retain moisture from the atmosphere — with the lower-average-molecular-weight PEGs capable of taking up much more moisture than the higher PEGs.[63] This inverse relationship between hygroscopicity and molecular weight occurs because the relative influence of the two hydroxyl groups on the polymer decreases as the polyether chain length increases.

This wide range of hygroscopicities makes the PEGs, particularly the lower ones, useful as plasticizers. Because of their hygroscopicities, the higher molecular weight PEGs are usually not employed as true humectants; instead, they modulate the moisture retention of liquids with which they are blended.

D. SOLUBILITY IN WATER

As a result of their multiple ether linkages and two primary hydroxyl groups, both the liquid and solid PEGs are readily soluble in water.[87] The water solubilities of the PEGs decrease with increasing molecular weight.[63] For this reason, the solid PEGs may have to be heated to increase the rate of dissolution. Combined with their solubility in aromatic compounds, this water solubility has led to broad PEG utilization as vehicles or dispersants of dyes and medicines.[63]

E. SOLUBILITY IN ORGANIC SOLVENTS

The PEG family offers a range of solubilities in organic solvents. Like their solubilities in water, solubility of the PEGs in organic solvents decreases with increasing average molecular weight.[63] Polar organic solvents such as alcohols, glycols, ketones, and esters can be used to dissolve the PEGs, but the PEGs are insoluble in low-polarity solvents like aliphatic hydrocarbons. PEG solubility in liquids is determined by the chemical groups present in the would-be solvents. The ability of chemical groups to solvate the PEGs follows the order (listed in decreasing solvating efficacy): hydroxyl, halogen, aryl, ketone, ester, ether, and alkyl; compounds with ether and alkyl groups are essentially nonsolvents.[63]

F. PEG SOLVENCY

The PEGs are capable of dissolving many substances, especially highly polar ones. The higher molecular weight PEGs are more able to solubilize water-insoluble compounds than the lower PEGs.[63] This property makes the higher PEGs especially useful in the pharmaceutical industry because many medicants can be dissolved in them and, in turn, the water solubility of the PEGs allows the medicants to be readily absorbed by the primarily water-based human body.[63]

G. ADDITIONAL PROPERTIES

As can be seen from the discussions of viscosity, freezing/melting ranges, etc., the molecular weight of PEG has significant effects on its properties. This relationship manifests itself in properties other than those discussed above. For example, as the molecular weight of the PEGs increases, so do the specific gravities and flash points; vapor pressures, however, decrease with increasing molecular weight.[63]

In addition, the PEGs feature good stability except at high temperatures or when exposed to air, which may lead to some slight oxidation. Also, they are not volatile compounds, although PEG 200 and PEG 400 may be very slightly volatile because they contain small fractions of the simple glycols. Finally, the PEGs are essentially nontoxic, allowing them to be used by the cosmetic and food industries.

H. BLENDED PEG COMPOUNDS

In cases when the diverse properties offered by the available PEG products are insufficient for a specific application, two or more different PEGs can be blended to create an intermediate compound with intermediate properties. An example of the compounds that can be produced by this blending process is Union Carbide®'s Carbowax®* 540 Blend, a mixture of PEG 1450 and PEG 300.[87] For added compound diversity, the PEGs can also be blended with water, other glycols, or alcohols.[63]

* Registered trademark of Union Carbide® Corporation, Danbury, CT.

III. SYNTHESIS OF THE PEGs

A. HISTORICAL SYNTHESIS METHODS

In the past, the PEGs have been synthesized by a number of methods. Over 130 years ago, both Lourenco and Wurtz were able to prepare PEGs by heating ethylene glycol with water.[28] In 1939, Fordyce et al. produced 6-, 18-, and 42-membered PEGs by reacting a dichloride of a PEG with 2 mol of a monometallic salt of a PEG. The monometallic salt had to be used to avoid further reactivity with the reaction products, which could have produced PEGs of indefinite length.[28]

$$\underset{\text{(a dichloride of a PEG)}}{ClCH_2CH_2O{-}(CH_2{-}CH_2{-}O)_x{-}CH_2CH_2Cl} + 2\ \underset{\text{(monometallic salt of a PEG)}}{KOCH_2CH_2O{-}(CH_2{-}CH_2{-}O)_y{-}CH_2CH_2OH} \rightarrow$$

$$\underset{\text{(a higher PEG)}}{HOCH_2CH_2O{-}(CH_2{-}CH_2{-}O)_{x+2y+4}{-}CH_2CH_2OH} + 2\ KCl \qquad (1)$$

B. CURRENT INDUSTRIAL SYNTHESIS METHODS

These early synthesis methods are not typically used in the commercial production of polyethylene glycols. Instead, the PEGs are prepared by the stepwise reaction of ethylene oxide with a growing polyether chain, each step adding another oxyethylene group ($-CH_2-CH_2-O-$).

1. Synthesis-Reaction Theory

A simplified version of this stepwise addition can be seen in the production of the two simplest ethylene glycols. (In this example, only reaction 3 is truly representative of the commercial PEG-synthesis procedure, the first step being included only to help illustrate the stepwise addition process.) In the first step, ethylene oxide is reacted with water, producing ethylene glycol:

$$\underset{\diagdown\ O\ \diagup}{H_2C{-}CH_2} + HOH \longrightarrow HOCH_2{-}CH_2OH \qquad (2)$$

In the second step, another ethylene oxide molecule is added. It reacts with the ethylene glycol produced in reaction 2, resulting in the formation of diethylene glycol:

$$\underset{\diagdown\ O\ \diagup}{H_2C{-}CH_2} + HOCH_2{-}CH_2OH \longrightarrow HOCH_2CH_2{-}O{-}CH_2CH_2OH \qquad (3)$$

This reaction continues adding oxyethylene groups to the growing polymer until the desired molecular-weight PEG has been obtained. The general reaction for the preparation of PEGs can, therefore, be reduced to

$$HOH + {}_{n}C_2H_4O \longrightarrow HO\text{–}(C_2H_4O)_n\text{–}H \qquad (4)$$

where $n > 3$ (i.e., diethylene glycol and triethylene glycol are not considered true PEGs).

2. Industrial-Reaction Conditions

In industrial production of the lower-average molecular weight PEGs, reaction 2 or both reactions 2 and 3 are omitted. Instead, ethylene glycol or diethylene glycol is heated in an autoclave in the presence of a caustic catalyst, and ethylene oxide is gradually added until the desired molecular weight PEG has been obtained.[56] To speed up the reaction rate, an alkaline catalyst such as sodium hydroxide is typically added, a pressure of 20 to 50 psig is maintained, and the reaction is heated to 150 to 180°C.[56] After neutralization of the catalyst, the resulting PEG is decolorized and (sometimes) filtered. In the production of the higher PEGs, (e.g., PEG 4000), this same stepwise reaction is utilized, but a higher-molecular-weight PEG, such as PEG 600, replaces the ethylene glycol in the autoclave at the start of the reaction.[56]

The ethylene glycols are produced in this manner for safety reasons. Only the amount of ethylene oxide required for the addition of one oxyethylene group is added at any one time. This helps control the highly exothermic reaction, which releases about 20 kcal for each mole of ethylene oxide that is reacted.[56]

IV. PEG DERIVATIVES

One of the most important applications of the PEGs is their use as chemical intermediates in the formation of derivatives. As previously indicated, the chemical reactivity of the PEGs results from the primary hydroxyl groups located at each end of the polymer. Fatty acids, fatty alcohols, or any hydrophobic compound that contains a reactive hydrogen can be reacted with one or both of these hydroxyl groups to create PEG derivatives. These derivatives include esters, salts, and ethers. The PEGs also react readily with dibasic acids to produce alkyd resins. An added benefit of PEG derivatives stems from the fact that derivatives with almost any hydrophobe/hydrophile balance can be created by simply altering the ratio of the reactants.[56]

A. PEG ESTERS

The most industrially significant of these derivatives are the PEG esters. Because of their low volatility, the higher-molecular-weight PEGs are esterified through an acid-catalyzed reaction with fatty acids. The production of these PEG mono- and diesters is a rather simple process.[87] Using an autoclave, esters are prepared by reacting a PEG with a fatty acid for 4 to 6 h. An acidic catalyst, such as sulfuric acid, should be used, and the reaction should be carried out under reduced pressure and at 120 to 130°C.

1. Applications for the PEG Esters

The PEG esters are used in diverse applications and by a variety of industries. The esters are most commonly used as water-soluble, surface-active compounds found in detergents.[63] They can also be found in herbicides,[46] corrosion inhibitors for oil wells,[12] and synthetic lubricants.[82] The PEGs can also be esterified with methacrylic acid to produce hard, colorless resins that can be used as surface coatings or to impregnate and laminate.[3]

The PEG esters themselves often serve as chemical intermediates. For example, the PEGs are combined with fatty acids, and the resulting PEG esters are subsequently reacted with an aliphatic amine to produce cost-efficient detergents, wetting agents, and emulsifying agents.[17]

B. OTHER PEG DERIVATIVES

In addition to the PEG esters, other PEG derivatives find their way into industrial products. For example, the PEGs are reacted with acryl halides to make ethers that can be used as plasticizers.[22] Since the late 1950s, they also have been reacted with alkyd resins to produce stable, water-dispersible alkyd resins for high-gloss paints.[5] Alternatively, by reacting the PEGs with a polybasic carboxylic acid and subsequently, with a water-insoluble acid, water-soluble demulsifiers are produced for the oil industry.[20] Elastomer sponges have also been created by reacting PEGs with arylene diisocyanate.[39]

V. INDUSTRIAL APPLICATIONS FOR THE PEGs

Reference 76 provides an extensive review of applications for the PEGs and their derivatives. Additional information on PEG applications can be obtained from the technical bulletins available from large-scale PEG suppliers, such as Union Carbide® Corporation and the Dow® Chemical Company.

A. PHARMACEUTICALS

The PEGs are widely used by the pharmaceutical industry because of their solvency action, water solubility, and blandness. The PEGs are commonly utilized in the production of solid dispersions of drugs;[27,34,38,70] because of the PEGs' solvating action, water-insoluble drugs can be dissolved in PEG, and, subsequently, the PEGs' water solubility allows these drugs to be dispersed in the body. The PEGs are also used in the preparation of suppositories[71] and sustained-release tablets.[24,50]

B. MEDICINE

PEGs are frequently used in medical products and studies. For example, they are used to coat surfaces in hopes of preventing platelet adhesion and subsequent clotting.[84] They have also been utilized as pliable, nonadherent films for surgical dressings for application to open lesions.[59,88] The PEGs' inertness, water solubility, and the fact that they are composed of molecules of different sizes that pass through membranes at different speeds have allowed the PEGs to serve in studies of intestinal permeability.[14]

C. COATINGS, PAINTS, AND INKS

Water solubility, pigment-binding capability, hygroscopicity, stability, and spreading ability all make the PEGs useful in surface coatings, paints, and inks.[63] For example, the PEGs are used to make hard, impact-resistant high-solid thermosetting coating compounds[49] and also serve as vehicles for the application of coatings for glass or metal.[67] In the treatment of paints, the PEGs give alkyd resins water dispersibility[5] and modify paints (e.g., metallic automotive enamels) so that they won't sag or pull away from the edges.[9] The PEGs are also utilized in the manufacture of water-resistant, air-drying paints.[74] PEGs are often added to inks to minimize hygroscopicities, as well as to jet-printing inks[55] and inks found in pressure-release decals,[43] where they increase stability.

D. COSMETICS, TOILETRIES, AND HOUSEHOLD PRODUCTS

The PEGs are common in cosmetics and toiletries because of their blandness, water solubility, and solvency action.[63] In addition, they add stability to cosmetic products and do not produce the greasy feeling found in some other cosmetics.[76] Because of all these qualities, the PEGs can be found in toothpastes,[64] hydrophilic denture adhesives,[21] antiperspirants,[61,86] perfumes,[91] and cosmetic regimens.[58]

The PEGs and their ester derivatives are also found in household products because of their water solubility, surface activity, stability, and soil-repelling nature. These qualities make them useful in fabric conditioners,[42] bleaching compounds,[69] and detergents,[25] as well as in cleaners for windows, glass, and chrome.[16]

E. PETROLEUM CHEMICALS AND CERAMICS

In the petroleum industry, PEGs are utilized for the treatment of both products and the oil wells used to obtain these products. For example, PEGs are employed to prepare corrosion inhibitors for use in oil wells,[12,13,15] but they are also used in the treatment of distillate-type fuels[19] and asphalt.[33]

Water-solubility and lubricity make the PEGs useful in the ceramic industry, where they serve as vehicles for ceramic pigments and coatings.[40,66] Their ability to burn off without excessive carbon buildup also makes the PEGs useful in high-temperature ceramic molding.[63]

F. OTHER INDUSTRIAL APPLICATIONS

In addition to the general application groups discussed previously, the PEGs are well suited for use in a number of other products and processes because of their physical properties. They are utilized in the production of lubricants[68] and wax casting molds[54,57] because their water-solubility allows them to be washed away easily after processing. The PEGs are also commonly used in the treatment of paper,[4,83] plastics,[37] rubber compounds,[57] leather,[29] and textiles.[18,81,89]

VI. SCIENTIFIC APPLICATIONS FOR PEG

Because many different PEGs can be used in the same scientific application, the PEGs will be referred to as a single entity in this section.

In addition to its use as an embedding medium in immunohistochemistry (IHC) and electron microscopy (EM), PEG has numerous applications in the scientific world. Molecular biology, immunology, enzymology, and protein chemistry are just some of the scientific disciplines that take advantage of the unique properties of PEG.

A. MOLECULAR BIOLOGY APPLICATIONS

There are two major types of molecular biology applications in which PEG is used. The first type is made up of those applications that exploit PEG for its macromolecular-crowding capability. The second type comprises purification methods in which specific nucleic acids or bacteriophage are purified with PEG by virtue of their size.

1. Macromolecular-Crowding Applications

The ability of PEGs to simply occupy volume in solutions is often exploited in molecular biology. By simply being present in a reaction mixture, PEG excludes reaction components from a large portion of the solution. This results in a higher effective concentration of other reaction components in the regions of the reaction mixture where PEG is not present.

An example of this can be seen when using single-stranded probes (RNA or cDNA) to perform nucleic acid hybridization on a filter. PEG greatly increases the hybridization rate because the effective probe concentration in solution is higher, allowing it to locate and bind to its target more rapidly.[2] While other polymers, such as dextran sulfate, are also used in this application, they are not as effective as PEG at accelerating the hybridization rate; the required 10% PEG solution is also easier to handle because of its lower viscosity. This use of PEG to accelerate the rate of probe hybridization also holds true when probing with single-stranded probes that have been cross-linked to proteins useful in colorimetric detection of DNA, such as peroxidase.[78]

The macromolecular-crowding ability of PEGs manifests itself in other applications in the field of molecular biology. For example, PEG accelerates the rate of T4 DNA ligase-mediated ligation of DNA by increasing the relative concentration of DNA fragments in a solution so that they are more likely to encounter each other.[72,92] Similarly, λ DNA fragments that have complementary ends are "crowded", accelerating their nonenzymatic ligation.[36] This same macromolecular-crowding process may also explain why the addition of PEG to T4 polynucleotide kinase-catalyzed reactions resulted in greatly increased amounts of product; perhaps by reducing the available volume in a solution, PEG increases the effective concentration of the reaction components.[36]

2. Purification Applications

PEG is often used in molecular biology purification procedures. In these applications, PEG is used to purify nucleic acids or bacteriophage on the basis of their size. An example of this is the use of PEG to purify bacteriophages from crude lysates of infected bacteria.[90] This procedure yields solutions that are 100 times more concentrated in phage than the original lysate.

PEG can also be used for the size fractionation of DNA fragments.[53] This application is based on the fact that a specific PEG concentration will selectively precipitate all DNA greater than a certain size. By gradually increasing the PEG concentration, smaller and smaller pieces of DNA can be incrementally precipitated out of solution.

Using this same method, PEG will separate other DNAs on the basis of size. For mammalian-transfection or molecular-cloning experiments, PEG has been used to purify bacterial plasmid DNA from bacterial host DNA; Sambrook et al. provide a protocol for purifying plasmids in this way.[79] PEG has also been utilized to separate undesirable DNA fragments from a vector in cloning experiments.[23]

B. CELL-FUSION APPLICATIONS

PEG is a known fusogen; that is, it is able to induce cell fusion between two cell types or between bacterial protoplasts and mammalian cells. For this reason, immunologists, somatic cell researchers, and molecular biologists use it in several routine applications.

For a complete review of cell fusion, see Reference 45. For cell-fusion procedures, see Reference 35; in that reference, Harlow and Lane (1988) provide two protocols for PEG-induced fusion, one that involves stirring cells with 50% PEG, and another that calls for centrifugation of the two cell types with 30% PEG. They also note that the use of PEG 1500 for cell fusion is most common, although any PEG from 1000 to 6000 will work.

1. Fusion of Mammalian Somatic Cells

While PEG had been used in the fusion of higher plant protoplasts (plant cells with cell walls that have been hydrolyzed by enzymes), Pontecorvo (1975) was the first researcher to use PEG to induce fusion of mammalian cells in monolayers, producing hamster/mouse somatic cell hybrids.[75] Now PEG is commonly used by somatic cell biologists to stimulate the fusion of cells from two different mammals. By adding dimethylsulfoxide as well as PEG, different mammalian cell types in suspension can also be fused.[32]

The exact method of PEG-induced membrane fusion is not clearly known. It is speculated that, in the case of phospholipid vesicle-vesicle fusion, lower concentrations of PEG (5% PEG 8000) first promote aggregation of vesicles, but higher PEG concentrations (35%) are required for actual fusion.[65] Fusion may occur because PEG reduces the dielectric constant of water, possibly destabilizing the membrane to such a degree that aggregated membranes are able to fuse.[6]

2. Protoplast/Mammalian Cell Fusion for DNA Transfer

One method of introducing cloned DNA into mammalian cells is to fuse bacterial protoplasts (lysozyme-treated bacteria) with mammalian cell cultures. The ability of PEG to enhance cell fusion makes it useful in these procedures. For example, PEG-induced fusion of bacterial protoplasts and mammalian cells in culture has been shown to produce an increased frequency of gene transfer.[77] When optimized, this technique results in a 10- to 20-fold increase in the frequency of direct transfer of cloned genes from bacterial protoplasts to mammalian cells.[80]

3. Hybridoma Production

The production of hybridomas, permanent cell lines secreting antibody directed against a specific antigen, often relies on PEG-mediated cell fusion. Köhler and Milstein (1975) first developed a method for fusing myeloma cells with spleen cells for the production of hybridomas.[48] Galfre et al. (1977) then adapted this technique by using PEG to increase the efficiency of cell fusion.[31]

C. PRODUCTION OF PEG-MODIFIED PROTEINS

PEG is often used to modify proteins in an attempt to produce proteins with more desirable characteristics. PEG-modified proteins are commonly employed in two application areas. In the first area, PEG molecules are bound to enzymes, making the enzymes soluble in organic solvents so that they can catalyze reactions not possible in an aqueous environment. The second area comprises applications in which proteins are modified with PEG to increase their therapeutic value.

1. Enzyme Modification for Solubility in Organic Solvents

Even though enzymes are supposed to be able to catalyze reactions in both directions, the excess of water molecules in an aqueous environment always drives a reaction in one direction (e.g., in the direction of hydrolysis when using enzymes such as chymotrypsin). In organic solvents, however, the reverse reaction is possible and, often, probable.[41] Unfortunately, most hydrolytic enzymes are insoluble in organic solvents.

This can be changed by enzyme modification with PEG. As indicated earlier, PEG is soluble in polar organic solvents; therefore, by attaching enough PEG molecules to an enzyme, a newly modified enzyme that is soluble in polar organic solvents can be created. Typically, these PEG molecules are added by reacting a PEG derivative such as 2,4 bis (*o*-methoxypolyethylene glycol)-6-chloro-triazine (often referred to as ''activated PEG_2'') with the enzyme. This newly modified enzyme can then be used to catalyze the reverse reaction in polar organic solvents.[41]

Typical examples of this enzyme-modification process are the alteration of hydrolytic enzymes so that they are able to catalyze synthesis reactions.

For example, PEG-modified chymotrypsin has been used successfully in peptide synthesis in organic solvents.[30,62] Similarly, PEG_2-modified α-L-rhamnosidase is able to catalyze the synthesis of rhamnosides in polar organic solvents.[85] Synthesis reactions catalyzed by PEG-modified enzymes offer an additional benefit: unlike reactions carried out in aqueous solutions, only a very small percentage of the newly synthesized product is hydrolyzed immediately after synthesis, resulting in higher yields of synthesis.[30]

2. PEG-Modified Proteins for Therapeutic Use

PEG-modified proteins have been designed for the treatment of human diseases caused by congenital protein deficiencies. Modified proteins useful in this type of treatment would be comparable to the native protein in activity levels, would have a long circulating life (the amount of time after injection until the enzyme is cleared from the body), and would feature reduced immunogenicity so that the modified protein could be administered without stimulating the full immunological response elicited by the native enzyme.

Attempting to create an enzyme that would satisfy these criteria, Abuchowski et al. (1977) believed that a large, linear, hydrophilic molecule like PEG could be attached to an enzyme to form a "shell" around the enzyme. He hoped this shell would cover many of the enzyme's antigenic determinants, which would normally elicit immune response upon enzyme injection, but leave the enzyme accessible to the smaller substrates required for enzymatic activity.[1] To test this theory, he and his co-workers attached methoxypolyethylene glycols (PEG derivatives in which one of the two terminal hydroxyl groups has been replaced by a methyl ether group) to bovine liver catalase. They successfully created an enzyme that featured a high relative activity, triggered a reduced immunological response, and exhibited a longer circulating life.[1]

Other PEG-modified proteins have been created in hopes of treating deficiency-based diseases. For example, a PEG-modified version of superoxide dismutase that had increased stability and a longer circulating life was produced; unfortunately, its enzymatic activity level was too low for it to be used.[11] Similarly, Beauchamp et al. (1984) modified calf adenosine deaminase (ADA) with monomethoxypolyethylene glycol in hopes of treating people with an ADA deficiency, but the new enzyme maintained some of the immunogenicity of the native enzyme.[10] In a more successful case, increased circulation time was a significant benefit of modifying recombinant α_1-proteinase inhibitor with PEG, as modification increased the pharmacological efficacy of the protein in the treatment of emphysema patients who lack the native inhibitor.[60]

Some PEG-modified proteins will not only produce a reduced immune response upon injection, but will actually increase tolerance of an organism to the native antigen itself.[52] Atassi and Manshouri (1991) provide a procedure for using monomethoxypolyethylene glycols to create these "tolerogenic" conjugates of proteins.[7]

PEG-modified proteins may also prove to be useful in the treatment of allergies. By coupling PEG to antigen E (the major protein constituent of ragweed pollen), King et al. (1977) produced an immunogenic antigen that is able to induce immune response like the native antigen, but also features decreased allergenic activity, making it useful in immunotherapy.[47]

D. PROTEIN PRECIPITATION

PEG has been used for the fractional precipitation of proteins[8] and to induce crystal formation,[51] although the exact mechanisms by which these occur are not known. Possibly, proteins are precipitated according to the volume-exclusion theory, a theory that claims that PEG sterically excludes the protein from areas of the solvent until the resulting concentration of the protein exceeds its solubility, causing the protein to precipitate.[8] Another theory claims that, because PEG is excluded from the domain of a protein, a solution containing both PEG and a protein — particularly a highly hydrophilic protein — is thermodynamically unstable; in response to this destabilized atmosphere, PEG then precipitates the protein out of solution.[51]

REFERENCES

1. **Abuchowski, A., McCoy, J. R., Palczuk, N. C., van Es, T., and Davis, F. F.,** Effect of covalent attachment of polyethylene glycol on immunogenicity and circulating life of bovine liver catalase, *J. Biol. Chem.*, 252, 3582, 1977.
2. **Amasino, R. M.,** Acceleration of nucleic acid hybridization by polyethylene glycol, *Anal. Biochem.*, 152, 304, 1986.
3. **Anderson, A. W. and Rambler, E. O.,** U.S. Patent Application 671,395,633, 1950; Bifunctional methacrylates, *Chem. Abstr.*, 45, 3651, 1950.
4. **Anderson, W. E.,** U.S. Patent 2,776,913, 1957; Non-water repellant, high-slip paper, *Chem. Abstr.*, 51, 6156, 1957.
5. **Armitage, F. and Trace, L. G.,** Water-dispersible alkyds, *J. Oil Colour Chem. Assoc.*, 40, 849, 1957.
6. **Arnold, K., Herrmann, A., Pratsch, L., and Gawrisch, K.,** The dielectric properties of aqueous solutions of poly(ethylene glycol) and their influence on membrane structure, *Biochim. Biophys. Acta,* 815, 515, 1985.
7. **Atassi, M. Z. and Manshouri, T.,** Synthesis of tolerogenic monomethoxypolyethylene glycol and polyvinyl alcohol conjugates of peptides, *J. Protein Chem.*, 10, 623, 1991.
8. **Atha, D. H. and Ingham, K. C.,** Mechanism of precipitation of proteins by polyethylene glycols, *J. Biol. Chem.*, 256, 12108, 1981.
9. **Barsotti, R. J.,** U.S. Patent 4,411,951, 1983; High-solids coating containing silica and polyethylene glycol for rheology control, *Chem. Abstr.*, 100, 53282, 1983.
10. **Beauchamp, C., Dadonna, P. E., and Menapace, D. P.,** Properties of a novel PEG derivative of calf adenosine deaminase, *Adv. Exp. Med. Biol.*, 165, 45, 1984.
11. **Boccù, E., Velo, G. P., and Veronese, F. M.,** Pharmokinetic properties of polyethylene glycol derivatized superoxide dismutase, *Pharmacol. Res. Commun.*, 14, 113, 1982.
12. **Caldwell, J. A. and Lytle, M. L.,** U.S. Patent 2,614,982, 1952; Corrosion inhibitor for oil wells containing hydrogen sulfide, *Chem. Abstr.*, 47, 1922, 1952.

13. **Caldwell, J. A. and Lytle, M. L.,** U.S. Patent 2,799,649, 1957; Inhibitors for corrosion by petroleum-well fluids, *Chem. Abstr.*, 51, 15106, 1957.
14. **Chadwick, V. S., Phillips, S. F., and Hoffmann, A. F.,** Measurements of intestinal permeability using low molecular weight polyethylene glycols (PEG 400). I. Chemical analysis and biological properties of PEG 400, *Gastroenterology,* 73, 241, 1977.
15. **Chamot, W. M.,** U.S. Patent 2,824,059, 1958; Corrosion-inhibiting compositions for oil and gas wells, *Chem. Abstr.*, 52, 11400, 1958.
16. **Church, P. K.,** U.S. Patent 4,213,873, 1980; Water-based window, glass, and chrome cleaner, *Chem. Abstr.*, 93, 152049, 1980.
17. **Cook, A. A. and Sapers, I.,** U.S. Patent 2,491,478, 1949; Polyalcohol-fatty acid-aliphatic amine combinations (textile assistants), *Chem. Abstr.*, 44, 8130, 1949.
18. **Cook, A. A. and Sapers, I.,** U.S. Patent 2,593,413, 1952; Fatty acid-alkoxypolyglycol aliphatic amine combinations useful as textile surface agents, *Chem. Abstr.*, 46, 9330, 1952.
19. **Daigle, E. C.,** U.S. Patent 2,650,187, 1953; Color stabilization for fuel oils, *Chem. Abstr.*, 48, 2361, 1953.
20. **DeGroote, M. and Keiser, B.,** U.S. Patent 2,363,045, 1944; Demulsifiers, *Chem. Abstr.*, 39, 3909, 1944.
21. **Dhabhar, D. J. and Schmidt, N. F.,** U.S. Patent 4,530,942, 1985; CMC/PEG containing denture adhesive, *Chem. Abstr.*, 103, 165,949, 1985.
22. **Doelling, G. L. and Adams, K. H.,** U.S. Patent 2,527,506, 1950; Polyethylene glycol ethers, *Chem. Abstr.*, 45, 2970, 1950.
23. **Dunn, I. S. and Blattner, F. R.,** Charons 36 to 40: multi-enzyme, high capacity, recombinant deficient replacement vectors with polylinkers and polystuffers, *Nucleic Acids Res.*, 15, 2677, 1987.
24. **Dunn, J. M.,** U.S. Patent 4,520,009, 1985; Sustained-release aspirin formulation, *Chem. Abstr.*, 103, 42671, 1985.
25. **Ferry, M. W.,** U.S. Patent 4,276,205, 1981; Detergent compositions containing amine oxide, nonionic surfactants and polyethylene glycol, *Chem. Abstr.*, 95, 99703, 1981.
26. **Flory, P. J.,** Molecular size distribution in ethylene oxide polymers, *J. Am. Chem. Soc.*, 62, 1561, 1940.
27. **Ford, J. L. and Wallis, R. M.,** Importance of preparation conditions on some properties of polyethylene glycol 6000 (PEG 6000) solid dispersions, *J. Pharm. Pharmacol.*, 37 (Suppl.), 116P, 1985.
28. **Fordyce, R., Lovell, E. L., and Hibbert, H.,** Studies on reactions relating to carbohydrates and polysaccharides. LVI. The synthesis of the higher polyethylene glycols, *J. Am. Chem. Soc.*, 61, 1905, 1939.
29. **Fricke, E.,** New method for stuffing leather in tanning, *Chem. Abstr.*, 51, 7747, 1985.
30. **Gaertner, H. F. and Puigserver, A. J.,** Peptide synthesis catalyzed by polyethylene glycol-modified chymotrypsin in organic solvents, *Proteins: Structure, Function, Genetics,* 3, 130, 1988.
31. **Galfre, G., Howe, S. C., Milstein, C., Butcher, G. W., and Howard, J. C.,** Antibodies to major histocompatibility antigens produced by hybrid cell lines, *Nature,* 266, 550, 1977.
32. **Hales, A.,** A procedure for the fusion of cells in suspension by means of polyethylene glycol, *Somatic Cell Genetics,* 3, 227, 1977.
33. **Hardman, H. F.,** U.S. Patent 2,759,842, 1956; Asphalt compositions containing silica aerogels, *Chem. Abstr.*, 50, 17422, 1956.
34. **Hargreaves, B. J., Pearson, J. T., and Connor, P.,** Physical properties of betamethasone alcohol-polyethylene glycol 6000 solid dispersions, *J. Pharm. Pharmacol.*, 31 (Suppl.) 47P, 1979.
35. **Harlow, E. and Lane, D.,** *Antibodies: A Laboratory Manual,* Cold Spring Harbor Laboratory Press, Cold Spring Harbor, NY, 1988, chap. 6.

36. **Harrison, B. and Zimmerman, S. B.,** T4 polynucleotide kinase: macromolecular crowding increases the efficiency of reaction at DNA termini, *Anal. Biochem.*, 158, 307, 1986.
37. **Hassall, T.,** U.S. Patent 4,389,506, 1983; Poly(vinyl alcohol) dust suppression by admixing polyglycol, *Chem. Abstr.*, 99, 71703, 1983.
38. **Hassler, W. H. and Sperandio, G. J.,** The formulation of a water-soluble suppository base, *Chem. Abstr.*, 47, 5625, 1953.
39. **Hill, F. B., Jr.,** U.S. Patent 2,726,219, 1955; Polyalkylene ether glycol-arylene diisocyanate elastomer sponge, *Chem. Abstr.*, 50, 6830, 1955.
40. **Hoffman, L. C.,** U.S. Patent 2,823,138, 1958; Thermofluid vehicle for vitreous enamel color compositions, *Chem. Abstr.*, 52, 7646, 1958.
41. **Inada, Y., Nishimura, H., Takahashi, K., Yoshimoto, T., Saha, A. R., and Saito, Y.,** Ester synthesis catalyzed by polyethylene glycol-modified lipase in benzene, *Biochem. Biophys. Res. Commun.*, 122, 845, 1984.
42. **Ives, D. T., Madill, M. J., and Sliva, P. G.,** U.S. Patent 4,514,444, 1985; Fabric cleaning/conditioning compositions, *Chem. Abstr.*, 103, 72892, 1985.
43. **Johnson, R. E.,** U.S. Patent 4,261,749, 1981; Thermoplastic inks suitable for pressure-release type decalcomania, *Chem. Abstr.*, 95, 9103, 1981.
44. **Kastens, A. S.,** Polymers from 1,2-epoxides: applications and technology, in *Polyethers,* Part 1; *High Polymers,* Vol. 13, Mark, H. et al., Eds., Interscience, New York, 1963, 169.
45. **Kennet, R. H.,** Cell fusion, *Methods Enzymol.*, Jakoby, W. B. and Pastan, I. H., Eds., 58, 345, 1979.
46. **King, L. J. and Kramer, J. F.,** Herbicidal properties and volatility of polyethylene glycol and polypropylene glycol esters of 2,4-D and 2,4,5-T, *Chem. Abstr.*, 46, 11549, 1951.
47. **King, T. P., Kochoumian, L., and Lichtenstein, L. M.,** Preparation and immunochemical properties of methoxypolyethylene glycol coupled- and N-carboxy methylated-derivatives of ragweed pollen allergen antigen E, *Arch. Biochem. Biophys.*, 178, 442, 1977.
48. **Köhler, G. and Milstein, C.,** Continuous cultures of fused cells secreting antibody of predefined specificity, *Nature,* 256, 495, 1975.
49. **Kuhn, R. R., Whiteman, J. D., and Brendley, W. H.,** U.S. Patent 4,515,835, 1985; High-solids thermosetting coating compositions, *Chem. Abstr.*, 104, 226436, 1985.
50. **Landahl, A. R. and Erlandson, S.,** U.S. Patent 4,557,925, 1985; Membrane-controlled sustained release tablets and method, *Chem. Abstr.*, 104, 213250, 1985.
51. **Lee, J. C. and Lee, L. L. Y.,** Preferential solvent interactions between proteins and polyethylene glycols, *J. Biol. Chem.*, 256, 625, 1981.
52. **Lee, W. Y. and Sohen, A. H.,** Abrogation of reaginic antibodies with modified allergens, *Nature,* 267, 618, 1977.
53. **Lis, J. T.,** Fractionation of DNA fragments by polyethylene glycol induced precipitation, *Methods Enzymol.*, Grossman, L. and Moldave, K., Eds., 65, 347, 1980.
54. **Livengood, S. M.,** U.S. Patent 2,567,076, 1951; Polyethylene glycol superesters, *Chem. Abstr.*, 45, 10675, 1951.
55. **Makashima, H., Toyoda, T., Okamura, N., Yano, H., Mizoguchi, A., and Hiromori, Y.,** U.S. Patent 4,279,653, 1981; Ink compositions for ink jet recording, *Chem. Abstr.*, 95, 117260, 1981.
56. **Malkemus, J. D.,** Production of alkylene oxide derivatives, *J. Am. Oil Chem. Soc.*, 33, 571, 1956.
57. **Marks, B. M.,** U.S. Patent 2,468,094, 1949; Casting compositions comprising methyl methacrylate and glycol dimethylacrylates, *Chem. Abstr.*, 43, 6467, 1949.
58. **Martin, J. O.,** U.S. Patent 4,268,502, 1981; Cosmetic tonic formulation, *Chem. Abstr.*, 95, 30231, 1981.
59. **Masci, J. N. and George, W. L.,** U.S. Patent 2,772,999, 1956; Hemostatic surgical dressing, *Chem. Abstr.*, 51, 3094, 1956.

60. **Mast, A. E., Salvesen, G., Schnebli, H.-P., and Pizzo, S. V.,** Evaluation of the rapid plasma elimination of recombinant α_1-proteinase inhibitor: synthesis of polyethylene glycol conjugates with improved therapeutic potential, *J. Lab. Clin. Med.*, 116, 58, 1990.
61. **Mast, R.,** U.S. Patent 4,338,294, 1982; Antiperspirant compositions, *Chem. Abstr.*, 97, 115170, 1982.
62. **Matsushima, A., Okada, M., and Inada, Y.,** Chymotrypsin modified with polyethylene glycol catalyzes peptide synthesis reaction in benzene, *FEBS Lett.*, 178, 275, 1984.
63. **McClelland, C. P. and Bateman, R. L.,** Technology of the polyethylene glycols and Carbowax® compounds, *Chem. Eng. News,* 23, 247, 1945.
64. **Miles, J. J. and Demasi, D. F.,** U.S. Patent 4,599,363, 1986; Wetting and dispersing powders, *Chem. Abstr.*, 105, 158654, 1986.
65. **Morgan, C. G., Thomas, E. W., and Yianni, Y. P.,** The use of fluorescence energy transfer to distinguish between poly(ethylene glycol)-induced aggregation and fusion of phospholipids, *Biochim. Biophys. Acta,* 728, 356, 1983.
66. **Morris, W. C.,** U.S. Patent 2,617,740, 1952; Vehicle for screen-process ceramic coating, *Chem. Abstr.*, 47, 3538, 1952.
67. **Morris, W. C.,** U.S. Patent 2,748,093, 1956; Screen-process vehicles for application of coatings to glass or metal, *Chem. Abstr.*, 50, 12502, 1956.
68. **Mueller, E. R. and Martin, W. H.,** Polyalkylene glycol lubricants. Uniquely water-soluble, *Chem. Abstr.*, 83, 166509, 1975.
69. **Murphy, A. P., Kassamali, A., and Curry, J. D.,** U.S. Patent 4,486,327, 1984; Bodies containing stabilized bleach activators, *Chem. Abstr.*, 102, 64018, 1984.
70. **Niazi, S.,** Effect of polyethylene glycol 4000 on dissolution properties of sulfathiazole polymorphs, *J. Pharmac. Sci.*, 65, 302, 1976.
71. **Palmieri, A.,** Dissolution of suppositories V: influence of aging on aspirin release from polyethylene glycol suppositories with and without crospovidone, *Chem. Abstr.*, 105, 178318, 1986.
72. **Pheiffer, B. H. and Zimmerman, S. B.,** Polymer-stimulated ligation: enhanced blunt- or cohesive-end ligation of DNA or deoxyribonucleotides by T4 DNA ligase in polymer solutions, *Nucleic Acids Res.*, 11, 7853, 1983.
73. Polyethylene glycol, in *The Merck Index: An Encyclopedia of Chemicals, Drugs, and Biologicals,* 10th ed., Windholz, M., Ed., Merck & Co., Rahway, NJ, 7439, 1983.
74. **Pons, D. A. and Bijker, A.,** U.S. Patent 4,477,623, 1984; Acrylic emulsion paints, *Chem. Abstr.*, 102, 26505, 1984.
75. **Pontecorvo, G.,** Production of mammalian somatic cell hybrids by means of polyethylene glycol treatment, *Somatic Cell Genetics,* 1, 397, 1975.
76. **Powell, G. M., III,** Polyethylene glycol, in *Handbook of Water-Soluble Gums and Resins,* Davidson, R. L., Ed., McGraw-Hill, New York, 1980, chap. 18.
77. **Rassoulzadegan, M., Binetruy, B., and Cuzin, F.,** High frequency of gene transfer after fusion between bacteria and eukaryotic cells, *Nature,* 295, 257, 1982.
78. **Renz, M. and Kurz, C.,** A colorimetric method for DNA hybridization, *Nucleic Acids Res.*, 12, 3435, 1984.
79. **Sambrook, J., Fritsch, E. F., and Maniatis, T.,** *Molecular Cloning: A Laboratory Manual,* 2nd ed., Cold Spring Harbor Laboratory Press, Cold Spring Harbor, NY, 1989, 1.40.
80. **Schaffner, W.,** Direct transfer of cloned genes from bacteria to mammalian cells, *Proc. Natl. Acad. Sci. U.S.A.*, 77, 2163, 1980.
81. **Seymour, R. B. and Schroder, G. M.,** U.S. Patent 2,486,804, 1949; Coating cellulosic textiles with maleic anhydride-styrene copolymer and solid polyethylene glycol, *Chem. Abstr.*, 44, 3263, 1949.
82. **Smith, P. V., Jr.,** U.S. Patent 2,575,195, 1951; Mixed esters of polyhydric alcohols and dibasic acids, *Chem. Abstr.*, 46, 1760.
83. **Stamm, A. J. and Cohen, W. E.,** Swelling and dimensional control of paper. I. Effect of wet-strength resins and Carbowax®, *Chem. Abstr.*, 51, 9155, 1956.

84. **Tseng, Y. C. and Park, K.,** Synthesis of photoreactive poly(ethylene glycol) and its application to the prevention of surface-induced platelet activation, *J. Biomed. Mater. Res.*, 26, 373, 1992.
85. **Turecek, P. L., Kopf, A., and Pittner, F.,** Chemical modification of α-L-rhamnosidase with activated polyethylene glycol-synthesis method of various rhamnosides with different aglycones, *Biol. Chem. Hoppe-Seyler,* 369, 933, 1988.
86. **Turney, M. E.,** U.S. Patent 4,280,994, 1981; Antiperspirant stick compositions, *Chem. Abstr.*, 95, 138397, 1981.
87. **Union Carbide® Corporation,** *Carbowax® Polyethylene Glycols,* Union Carbide® Corporation, Danbury, CT, 1986.
88. **Ward, W. C.,** U.S. Patent 2,693,438, 1954; Preformed, nonadherent films for application to open lesions, *Chem. Abstr.*, 49, 2036, 1954.
89. **Wassell, H. E.,** U.S. Patent 2,393,865, 1946; Stain-removing composition, *Chem. Abstr.*, 40, 5170, 1946.
90. **Yamamoto, K. R., Alberts, B. M., Benzinger, R., Lawhorne, L., and Treiber, G.,** Rapid bacteriophage sedimentation in the presence of polyethylene glycol and its application to large-scale virus purification, *Virology,* 40, 734, 1970.
91. **Zeitz, V.,** U.S. Patent 4,185,086, 1980; Talc compositions, *Chem. Abstr.*, 92, 152,907, 1980.
92. **Zimmerman, S. B. and Harrison, B.,** Macromolecular crowding accelerates the cohesion of DNA fragments with complementary termini, *Nucleic Acids Res.*, 13, 2241, 1985.

Chapter 2

POLYETHYLENE GLYCOL-CELLOIDIN MIXTURE FOR HISTOLOGICAL AND HISTOCHEMICAL STUDIES

I. A. Efimov

TABLE OF CONTENTS

0-8493-4323-2/93/$0.00 + $.50

I. INTRODUCTION

It is very common to use the traditional immersing and embedding media celloidin, gelatin, or paraffin for histochemical research of animal tissues. These media have some shortcomings in practice that may limit the further study of the specimens obtained. Sometimes uses of the above-mentioned media are restricted because it is difficult to prepare sections with the required thickness or to obtain serial sections. Sometimes these substances have bad effects on the preservation of certain chemical components or biological activities of macromolecules.

In the past decades, new synthetic substances with the common name polyethylene glycols (PEGs) began to compete intensively with another immersing and embedding medium known as wax or paraffin. PEGs differ from each other in polymerization degree or their molecular weights and their melting temperatures. PEGs with molecular weights 1000 to 4000 become fluid at 37 to 60°C, respectively. PEGs have a common feature that is extremely valuable for morphologists and histochemists: all of them have good solubility in water or water-based solutions. Therefore, it is unnecessary to use alcohol or acetone for dehydration of biological specimens. In addition, one of the outstanding features of PEG is its ability to dissolve some organic compounds, for example, flaky celloidin. These features of PEGs provoked the creation of a new mixed PEG-celloidin embedding medium for histological and histochemical purposes.

I was first acquainted with the method for the treatment of biological specimens with PEG through a book written by Dr. Kissel that was published in Russian in 1962.[2] Unfortunately, this method did not become widespread because of large variability of its results, especially at a stage for obtaining and spreading sections. When celloidin is added to PEG, it becomes a stabilizing factor that decreases the highly hydrophilic nature of PEG and prevents sections from being destroyed. Taking advantage of good solubility of celloidin in PEG, we tried to combine the methods of embedding in celloidin and in PEG. Thus, a new mixture of immersing and embedding medium was produced, which retained the positive features of two known media but to some extent eliminated their defects. In this chapter the method for fixation of biological specimens, the composition of the new medium, and the characteristics and weight ratios of the components will be discussed. Practical preparations of the medium and some applications using PEG-celloidin embedding method for the study of histology with both vertebrate and invertebrate tissues will be described.

II. METHODS

A. EMBEDDING MEDIUM FOR OBTAINING 4- TO 12-μm SECTIONS

1. Composition of Embedding Medium

PEG 1500*	100 parts by weight
Flaky celloidin	10 parts by weight

In the summertime, PEG 3000 is recommended for easy sectioning. If moisture and temperature are high and section ribbon slips hardly along the microtome blade or adheres to it, this means the embedment is wet. To return it to a dry state, silica gel for thin layer chromatography must be added to the embedding medium 3 to 24 h before its use.

Proportions are:

Embedding medium	100 ml
Silikagel 15/40	10 g

Stir the mixture with a glass stick until it becomes its primary color.

2. Preparation of Embedding Medium

PEG of definite molecular weight should be placed into a glass beaker or china dish, then melted in a thermostatic oven. The melting point of PEG differs according to its molecular weight. The melting point of PEG 1000 is 37 to 39°C, and the melting points of PEGs with molecular weights from 1500 to 4000 are around 52 to 62°C. Usually, the melting point of PEG is labeled on the package. We purchased PEG from ''LOBA'' (Austria) or from various German firms.

Weighted celloidin is directly added to the melted PEG. Compressed clots of celloidin should be grounded before use. It takes 2 to 3 weeks for a complete solubilization of celloidin in PEG. After celloidin swells, it loses its white color and becomes invisible in melted PEG. It is necessary to slowly stir the mixture with a glass stick once a day. The mixture becomes homogenous and turns a pale yellow color within 2 to 3 weeks. (The higher the molecular weight of PEG used, the more intense the color of the mixture will be.) After that, the colored embedding medium is ready for use. The criterion of a well-prepared mixture is its brittleness when hardened. Elastic hardened medium is not suitable for the preparation of thin sections and it may be used only for preparing thick sections. Embedding medium that does not form thin celloidin film around the resulting section cannot be used for work.

* Average mol wt = 1500, the same expression as in other PEGs.

3. Storage of PEG

Molten PEG dissociates and converses gradually into lower molecular weight PEG. That is why the embedding medium may not be used for more than 3 to 6 months, when it is stored in molten form, even if it is kept in a closed container. The utilization time may be prolonged by the following treatment: some embedding medium is poured into a small glass vessel for immediate utilization, while the main part is stored in hardened form in a tightly capped glass container or polyethylene dish at room temperature.

4. Preparation of Immersing Medium

a. Immersing medium 1
   PEG 1000 -- 100 parts by weight
   Flaky celloidin -------------------------------------- 2 to 3 parts by weight
b. Immersing medium 2
   PEG 1500 -- 100 parts by weight
   Flaky celloidin --- 5 parts by weight

Immersing media 1 and 2 are prepared in a thermostatic oven at 56°C. Medium 1, ready for specimen immersing, is transferred into another oven at 37 to 39°C. The scheme for dissolving celloidin in PEG is similar to the procedure for preparation of embedding medium as previously described.

B. HISTOLOGICAL TREATMENT WITH PEG-CELLOIDIN IMMERSING AND EMBEDDING MEDIA

1. Fixation

Any fixatives with different compositions are suitable for the fixation of tissues for the processing of PEG-celloidin embedding, but those containing formalin are preferable. We used a fixative mixture composed of ethanol, formalin, and acetic acid.[3] To each 100 ml of the fixative, 5 ml of PEG 300 was added. The reason we used PEG 300 is explained as follows. First, PEG 300 penetrates easily through the fixed tissues, thus providing an equal distribution of the components of fixatives. Second, the presence of PEG 300 in fixed tissue may provide better chances for the safety of cells or tissues. Such fixative composition works mildly, hence fixation time must be increased 1.5- to 2-fold. PEG may be added only to fixing mixture that does not cause dimness and sediment.

2. Washing

The fixed specimen is transferred into 70% ethanol and then into distilled water. The duration of ethanol washing is about 1 h. If the fixing mixture contains acetic acid, washing should be continued until its smell disappears.

3. Dehydration of Fixed Specimens

The fixated specimen is transferred from water into PEG 300 with increasing concentrations: starting with 50% PEG in water, the sample is transferred to 70% of PEG in water, and finally to 100% of PEG. Dehydration time depends upon the size of the investigated specimen and upon the feasibility of the tissue for quick dehydration; the time varies from 2 to 24 h or longer. The criterion of dehydration is the velocity of specimen sinking to the bottom of the vessel. If the velocity is high, the specimens have to stay in PEG for about 2 h. A longer stay of the fixed material in PEG does not do too much harm to the cell or tissue. For specimens that lose water intensively, (i.e., fish eggs, etc.), then a mild dehydration procedure is adopted: starting from 5 to 10% PEG solution in water and increasing the PEG concentration gradually. (When fish eggs are treated, envelopes covering oocytes are not removed.)

Our experience showed that specimens might be dehydrated with a mixture of PEG 1000 and PEG 1500 in water solutions.

4. Transfer of Specimens in Immersing Medium

After specimens have reached the bottom of the vessel, they are transferred into PEG 1000 at 37 to 39°C in a thermostatic oven. The incubation time that the specimen has to spend in PEG (from 1 to 24 h) also depends on the properties of the specimen. This step is rather important because low molecular liquid PEG is being substituted by molten PEG of higher molecular weight during the incubation. Then the specimen is transferred into an immersing medium at 37 to 39°C for about 24 h. If the specimen is friable, it must stay in the immersing medium 1 for 2 to 3 days or longer.

Then the specimen is transferred into immersing medium 2 at 52 to 56°C. For histochemical revelation of enzymes, the specimen stays in the immersing medium for not more than 1 h. In all other cases the immersing time is 2 to 24 h. Chitinized animals (insects) are immersed for 3 days. Our experience indicated that if dehydration at the previous stages was good, the specimen could stay at immersing medium for 10 days without influencing their safety.

5. Embedding of Specimens

The embedding medium is poured into a watch glass (or any kind of similar device) for specimen embedding. A thin layer of glycerol is smeared at the inner surface of the bottom of the glass container. The specimen is transferred from embedding medium 2 to the bottom of the watch glass and oriented, if necessary. The embedding medium, together with the specimen, hardens at room temperature. Shortly after hardening of the embedding material, blocks of appropriate sizes are cut from the solidified embedment. This procedure will be difficult if it is carried out several hours after the hardening of the embedding medium, especially when PEGs 2000 to 4000 were used. If necessary, multiple reembedding of the specimen is possible.

6. Sectioning

The sectioning of the PEG-celloidin embedded tissue samples and the stickings up of the sections to the glass slides are quite similar to the procedures in paraffin embedding method. Usually, we use the glass slides pre-coated with protein-glycerol (1:1).[2,5]

7. Section Spreading

In contrast to paraffin section spreading, a heating table is not necessary when PEG-celloidin embedding medium is employed. If the microtome blade is well sharpened, then the section ribbon turns out evenly. Slices of section ribbon are transferred to the glass slide. Then 1 to 2 drops of hypertonic buffered solution or distilled waters are placed with a pipette by the edge of a section ribbon; the glass slide is tilted slightly, and these drops run along the sections, thus spreading them. The sections spread in this way are expanded by dissolving the PEG from the sections. A thin transparent celloidin film remains along the sections. The presence of celloidin film does not prevent visualization of specimens. However, this film of celloidin can be dissolved with an alcohol-ether mixture, if necessary, before embedding the specimen into mounting medium. This must be done with great care, because sections from the glass slide may detach after the celloidin film dissolves.

After the sections have been spread, they are dried and ready for further histological or histochemical procedures. Normally, we use a rapid specimen-drying procedure. When the sections are spread, the glass slide is placed carefully over the flame of an alcohol burner for a whole denaturation of proteins. Sections are dried quickly and are then ready for further treatment.

The sections spread in this way have one important advantage. Every tissue structure is seen well when moist spread sections are examined under the microscope, and the investigator can easily choose tissue sections that meet specifications for research.

8. Section Staining

Staining and any other histochemical procedures are performed on the slides after the dried sections are rinsed thoroughly with distilled water or a suitable buffer.

The stained sections are dehydrated with increasing concentrations of alcohol and cleaned with xylol, then they are mounted into Canadian resin for microscopic examination.

III. PRACTICAL USE OF PEG-CELLOIDIN METHOD IN HISTOLOGY AND HISTOCHEMISTRY

The potential application of this method to different research purposes is great, such as for morphological examination including impregnation with silver or sublimate to reveal enzyme in cells, histochemistry, and immunochemistry. Practical schemes of the given method for tissues of some animals both vertebrates and invertebrates, including chitinized larvae of insects, are given in the following section.

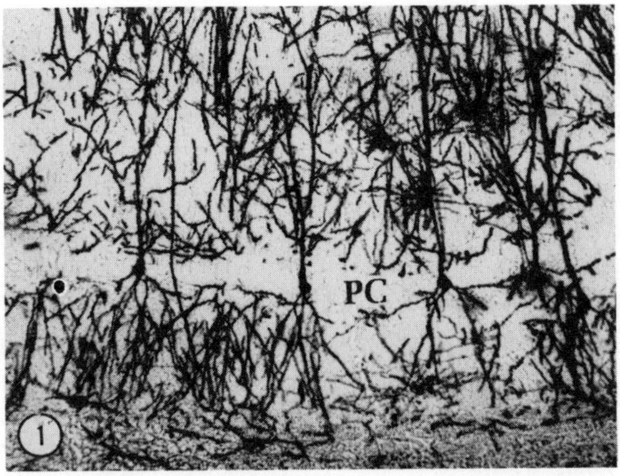

FIGURE 1. Impregnated rat brain according to Golgi-Cox method. PC, pyramidal cells of hippocampus; section thickness, 20 μm.

A. APPLIED TO THE STUDY OF VERTEBRATE TISSUES

As an example, the scheme of treatment of rat brain impregnated with sublimate is described as follows.

Rat brain is treated with sublimate according to Golgi-Cox method.[4] After the celloidin block with impregnated brain is obtained, or after it is cut into thick celloidin sections, a piece of tissue about 5 × 10 mm square and 3 to 4 mm thick (the piece may be somewhat larger, but then the time for treatment with each solution should be prolonged) is transferred into an aqueous solution of 70% PEG 300 for 2 h, then into 100% PEG 300 for another 2 h. The dehydrated specimen is then immersed into PEG 1000 and infiltrated for 24 h at 37 to 39°C for the displacing of liquid low molecular PEG by high molecular PEG. Then the piece is transferred into embedding medium (a mixture of 10% celloidin in PEG 1500) for 24 h at 56°C. The results of impregnation of 20 μm sections by Golgi-Cox method are presented in Figure 1. For obtaining thicker sections, specimens have to be embedded into embedding medium containing one-fourth PEG 1000. If any light structure appears in the center of the specimen (place the specimen on a glass plate for examination), this means that the impregnation was insufficient, and the specimen has to be returned into the embedding medium. Consequentially, the block must be prepared again.

For parenchyma organs of vertebrates, such as liver, spleen, and kidney, step by step impregnation is preferable — from the 2% celloidin in PEG 1000 at 37 to 39°C up to 5% celloidin in PEG 1500 at 55 to 56°C. Table 1 presents times required for the treatment of mouse kidney and guinea pig cerebellum by each solution for whole organ treatment. Here we would like to give out some practical advice that may aid the readers who wish to try this method. Generally, the method is based on a gradual substitution for the former solution

TABLE 1
Time for Fixation of Organs and Tissues and Subsequent Treatment

Specimen	Fixation	70% Ethanol	70% PEG 300 in water	100% PEG 300	100% PEG 1000	Immersing medium 1	100% PEG 1500	Immersing medium 2	Embedding medium
Mouse brain	24	2	Until specimen sinks to bottom of vessel	12	12	12	8	24	0.5
Guinea pig cerebellum	24	1	Same as above	10	10	10	8	24	0.5
Mouse kidney	24	1	Same as above	6	6	10	8	24	0.5
Rat brain impregnated	—	—	Same as above	12	12	—	12	24	0.5
Grain beetle larvae	24	2	Same as above	12	10	10	10	24	0.5
Planaria	12	1	Same as above	5	5	5	5	24	0.5
Leeches	24	1	Same as above	10	10	10	10	24	0.5

Note: (1) All time units of treatments are hour(s). (2) Guinea pig cerebellum, mouse brain, and kidney were infiltrated as a whole tissue. (3) If the histochemical method used has temperature limitations, then infiltration is held at 37 to 39°C, but time of treatment is increased.

already impregnated the specimen by a new one. The most important step of this method is the process of substitution of PEG 1000 at 37 to 39°C by immersing medium 1 containing 2% celloidin in PEG 1000. Impregnation time is determined quite simply. The mouse kidney is taken out of the immersing solution, cooled at room temperature, and cut in half with a scalpel. If there is any light spot in the center of it, the replacement of PEG 1000 by immersing mixture is incomplete. The specimen has to be returned to the immersing mixture. If the cutting surface is monochromatic without any feature of light inclusions, the kidney was immersed well. It is a sign that further treatment (i.e., immersion into 5% celloidin in PEG 1500) of the specimen becomes unnecessary. In this case, the specimen is rinsed in immersing medium for 5 to 10 min. Then the specimens are embedded into that medium. However, if the sections are cramped while the specimen is sectioned, then it must be transferred into 5% celloidin in PEG 1500 for the substitution of the previous medium for 24 h. Then the specimen is put into the immersing medium again. If this treatment does not prevent crumpling, immersing medium containing PEG 1500 must be replaced by one containing PEGs 2000 to 3000. Figures 2 to 4 are micrographs taken from PEG-celloidin sections of guinea pig cerebellum (5 μm), mouse kidney (7 μm), and pig thyroid gland, which were elaborated according to the previously mentioned procedures.

B. APPLIED TO THE STUDY OF INVERTEBRATE TISSUES; PARTICULARLY FOR CHITIN COVERED TISSUES

Our own experience showed the advisability of using PEG-celloidin embedding method for morphological and histochemical investigations in planaria, leeches, and the larvae of grain beetle. It is well known how difficult it is to study chitinized animals. Head ganglia, for instance, must be previously isolated from the animal body. Our method provides a possibility of avoiding this step. The head end together with two to three segments of the body are cut from the grain beetle larvae and are fixed in an appropriate solution and then treated by PEGs of increasing concentrations — from PEG 300 to PEG 1000 at 37 to 39°C. The real time of treatment by each solution is given in Table 1. It must be considered that if PEG is added to the fixing mixture (15 ml of PEG 300 to 100 ml of fixing mixture), then the piece may sink to the bottom of the vessel faster than indicated in Table 1. After the specimen has spent 4 to 12 h in PEG 1000 at 37 to 39°C, it is taken out and stuck to the smooth surface with the help of molten PEG. Afterwards, lateral parts of chitin and the front part of the head, together with the maxillae, are cut off. The specimen is transferred again into PEG 1000 and kept there for 12 h for a complete dehydration. Then the specimen is lowered into immersing medium 2 and made into blocks. If sections diffuse, the specimen must be returned right into the embedding medium for better impregnation and sections must be made again.

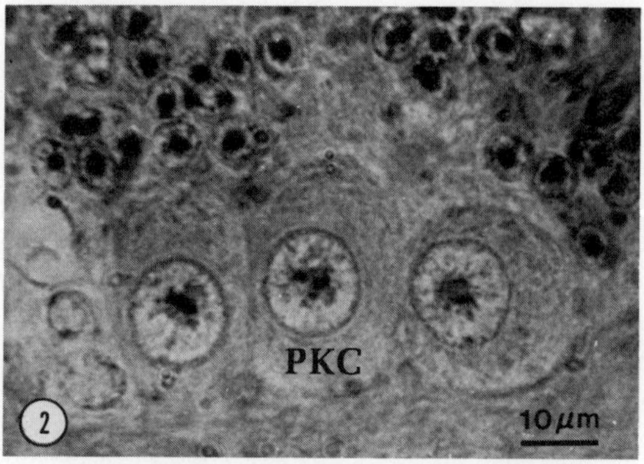

FIGURE 2. Guinea pig cerebellum, stained with hematoxilin-eosin. PKC, Purkinje's cells; section thickness, 4 μm, bar = 10 μm.

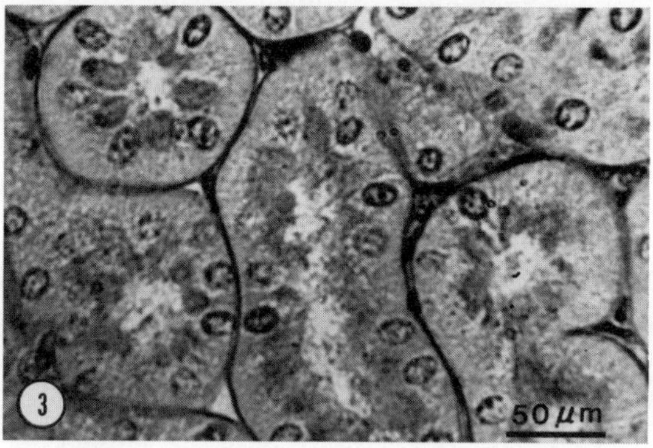

FIGURE 3. Mouse kidney, stained with hematoxilin-eosin. Section thickness, 7 μm; bar = 50 μm.

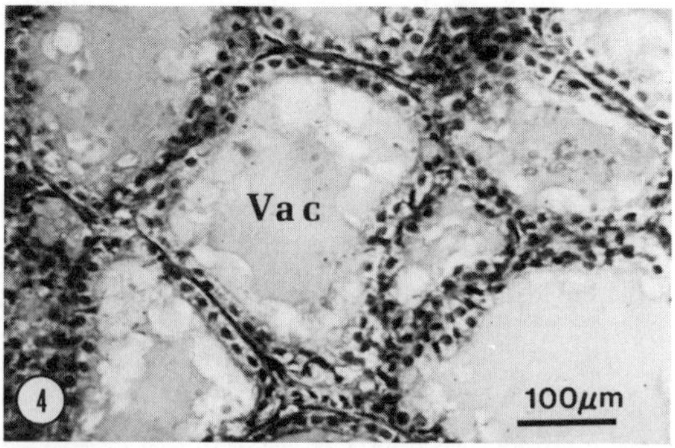

FIGURE 4. Pig thyroid gland (from Chernobyl). Vac, vacuolization in follicles (hyperfunction); bar = 100 μm.

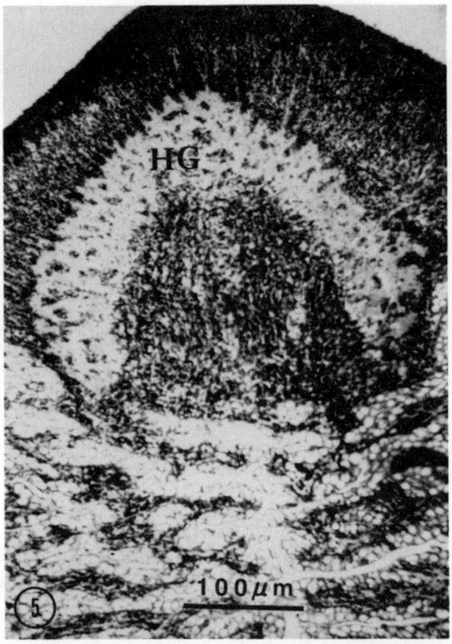

FIGURE 5. Head of planaria, stained with hematoxilin-eosin. HG, head ganglion of *Dugesia tigrina,* 5-μm section; bar = 100 μm.

Animals that live in water, such as planaria and leeches, dehydrate quite easily in PEG solutions and they are easily infiltrated. Taking planaria *Dugesia tigrina* and leech *Glossiphonia complanata* as examples, we demonstrated the possibility of using our method for water-living animals. The time needed for the treatment of these animals by each solution is listed in Table 1. Figures 5 and 6 are micrographs of sections of head ganglion of these animals as prepared with the PEG-celloidin embedding method. The section thickness is 5 μm.

The characteristic feature of planaria is that they are bending or twisting during fixation, so that it is very difficult to obtain flat sections along the whole body. To spread planaria before fixation, we used another method: several planaria were taken and transferred onto a glass slide together with water. Then equal amounts of 40% formalin and 2% Novocain® were mixed and warmed to 37 to 40°C. With the aid of a pipette several drops of this mixture were added near the planaria tail end. The planaria stretch and become flat.[1] By applying this procedure, planaria are fixed in solution suitable for experimentation.

C. APPLIED TO HISTOCHEMICAL INVESTIGATIONS

It is well known that the success of any histochemical reaction is directly related to the pre-treatment of the tissue sample before histochemical

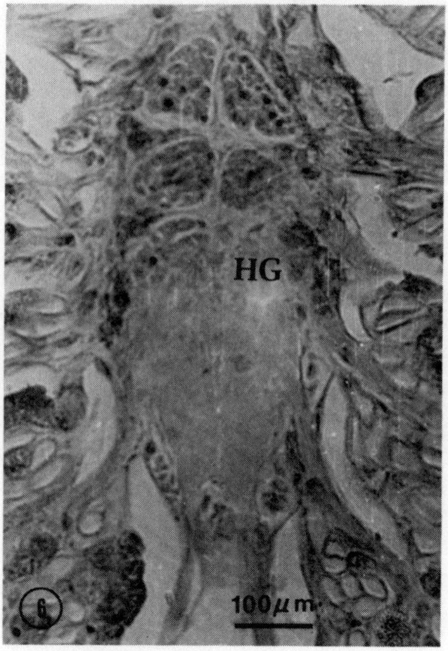

FIGURE 6. Head of leech *Glossiphonia complanata,* stained with hematoxilin-eosin. HG, head ganglion of leech; 5-μm section; bar = 100 μm.

preparation. Only with good preparation, can chemical structures of cells and tissues be properly preserved. If the PEG-celloidin method is used, the action of aggressive substances to destroy chemical structure of cells is greatly reduced. In cases where PEG-celloidin embedding method is applied, PEG not only serves as a component embedding the tissue, but also as a protector that preserves and stabilizes the chemical compounds of cells under dehydration.

All that has been referred to previously in this chapter has suggested that the scale of histochemical reaction on the tissue sections obtained by the PEG-celloidin embedding method is wide enough. Thus, a number of authors have modified this method and applied it successfully for revealing gangliosides of sea urchin by the indirect immunofluorescence method.[6,8] According to preliminary data reported by Sheiman et al.,[7] there is a possible immunocytochemical reaction for revealing endorphin in head ganglia cells of planaria *D. tigrina.* This method provides results quite well for discovering neurosecretion in neurosecretory cells of grain beetle larvae as shown in Figure 7. The neurosecretion was discovered according to the Paget method.

The widely accepted opinion is that a relatively high temperature (56°C) is unacceptable for some histochemical reactions and enzyme revealing. This is quite right if paraffin is taken for infiltration. In cases where our treatment is used, the temperature may be as high as 52 to 56°C.

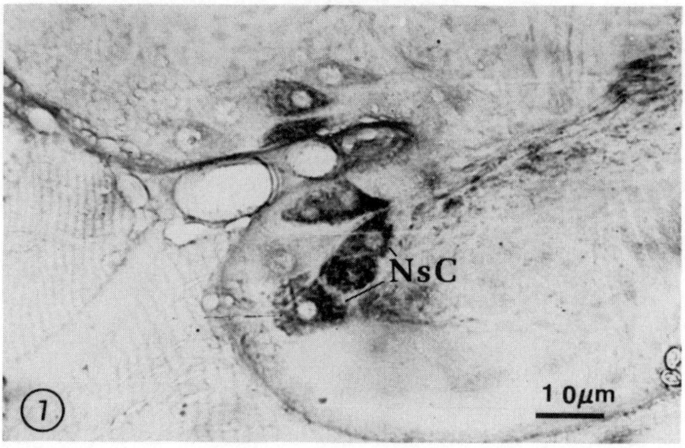

FIGURE 7. Ganglion of grain beetle larvae. Exposure of neurosecretion in neurosecretory cells. NsC, neurosecretory cells; 5-μm section; bar = 10 μm.

The aim of this chapter is not merely to report this method for morphological and histochemical investigation, but to get morphologists and histochemists to pay attention to this method. PEG-celloidin embedding method may still not quite familiar to the Western world; however, we suppose, it may be widely modified and adapted to different purposes by investigators.

IV. CONCLUDING REMARKS

A new infiltrating and embedding medium based on mixing PEG and celloidin was developed for histological and histochemical study of biological materials. Reliable procedures were established for the treatment animal tissues, including chitinized invertebrate samples. The new PEG-celloidin embedding techniques enable us to make sections of 3 to 10 μm thickness, preserving cellular structures and the chemical nature of the substances within the cells. This method is applicable to different systems according to the interests of researchers.

ACKNOWLEDGMENT

I greatly appreciate Dr. Kuixiong Gao for his kind invitation for me to take part in the creation of this book. Thanks also to Mr. Alvin W. Gao for the improvement of the structure of this chapter, and for transferring the typewritten form of the original manuscript to a diskette form.

REFERENCES

1. **Efimov, I. A. and Sakharova, N. Yu.,** A simple method of stretching planarians in the course of fixation, *Ontogenez,* 7, 537, 1976.
2. **Kisely, D.,** *Practical Microtechnic and Practical Histochemistry,* Hungarian Academy of Sciences, Budapest, 1962, 97 (in Russian).
3. **Lillie, R. D.,** *Histopathological Technic and Practical Histochemistry,* McGraw-Hill, New York, 1965, chap. 3.
4. **Ramon-Moliner, E.,** The Golgi-Cox technique, in *Contemporary Research Methods in Neuroanatomy,* Nauta, W. Y. H. and Ebesson, S. O. E., Eds., Springer-Verlag, New York, 1970, 32.
5. **Roskin, G. Y.,** *Microskopic Technic,* Sovetskaja Nauka, Moscow, 1957, 150 (in Russian).
6. **Sadykova, K. A., Zvezdina, N. D., Martynova, L. E., Guzareva, E. V., Racic, L., Rogac, L., and Buznicov, G. A.,** Gangliosides of the cell surface and cytoplasm in the early embryos of sea urchins under various influences, *Ontogenez,* 18, 413, 1987.
7. **Sheiman, I. M., Tiras, Kh. P., Vinogradov, V. A., and Efimov, I. A.,** Enkephalin analog dalargin — stimulates the planarian head regeneration, *Dokl. USSR Acad. Sci.,* 284, 481, 1985.
8. **Zvezdina, N. D., Sadykova, K. A., Mikailov, A. T., Buznikov, G. A., and Bergelson, L. D.,** Gangliosides of sea urchin embryos, their localization and participation in early development, *Eur. J. Biochem.,* 186, 189, 1989.

Chapter 3

POLYETHYLENE GLYCOL IN IMMUNOCYTOCHEMISTRY AND ENZYME HISTOCHEMISTRY

Jeanette Mowery

TABLE OF CONTENTS

0-8493-4323-2/93/$0.00 + $.50

I. INTRODUCTION: WHY PEG?

Immunocytochemistry (ICC) and enzyme histochemistry (EHC) are essential techniques in the cellular and subcellular localization of biological macromolecules in many areas of biomedical investigation. When using these techniques, the need to preserve tissue morphology must be balanced against the need to maintain antigenic reactivity or enzymatic activity of the molecule in question. The standard tissue preparation methods for light microscopy (LM) are paraffin embedding and sectioning and cryosectioning. Although paraffin sectioning provides good morphological preservation for LM, it has several disadvantages for ICC and EHC because it requires the use of harsh organic solvents and relatively high temperatures for embedding that often reduce or destroy antigen reactivity or enzymatic activity.[23] At best, an increased concentration of antibodies is necessary for antigenic or enzymatic localizations; at worst, the technique cannot be used because antigenicity or enzymatic activity has been destroyed. Frozen sections are generally useful for routine immunocytochemical work at the light level, since antigenic reactivity and enzymatic activity are well preserved by freezing. However, in some tissues,[1] morphology is not preserved sufficiently to precisely identify the location of antigen or enzyme. In addition, frozen sections cannot be used for electron microscopy (EM) without lengthy cryoprotection steps because the ultrastructure is destroyed by the freezing process. Therefore, it is difficult to correlate observations at the light microscope level with ultrastructural information. Polyethylene glycol (PEG) embedding represents a useful alternative to these two methods.

The use of PEG as an embedding medium was introduced several decades ago.[2,4] It was adopted by some investigators[9,10,17] because of its water solubility and lower infiltration temperatures, but it was not generally accepted, probably due to associated technical difficulties. Blocks of PEG-embedded tissues were hygroscopic and difficult to section, especially under humid conditions.[7,9] In addition, affixing tissue sections to slides required delicate handling procedures because the PEG support dissolved readily in buffer or other aqueous solution, thereby dispersing the section.

The increased use of ICC and EHC by many investigators, especially after the advent of monoclonal antibody technology,[11] prompted a search for alternative embedding methods and PEG was reevaluated.[1,15,23] Attracted by the possibility of milder embedding protocols due to lower melting points and the water solubility of PEG, these investigators have improved on PEG embedding protocols, and some creative solutions to the section-mounting problem have been devised. Because milder fixation can be used due to milder embedding conditions, and the sections can be post-fixed and reembedded for EM, the PEG embedding method can be an excellent choice for ICC and IHC.

In this chapter, the use of PEG in ICC and EHC will be reviewed, and technical procedures associated with some of these methods will be discussed. Results of the author's work with PEG 600 embedding[16] will be presented.

Although the focus here is on ICC and EHC, it should be noted that PEG embedding is also useful for visualizing histofluorescence of catecholamines[22] and the staining patterns of injectable dyes, such as lucifer yellow.[23]

II. CHOICE OF PEG MOLECULAR WEIGHT FOR EMBEDDING

An important advantage to PEG embedding is the fact that lower tissue infiltration temperatures are possible, because of the lower melting points of PEG relative to paraffin embedding. Several different molecular weights of PEG have been used for ICC and EHC by different investigators as shown in Table 1. The reader is referred to the individual papers for exact experimental details and results. Some investigators found that the morphology of PEG-embeded tissue sections was superior to that of frozen sections and comparable to that of paraffin embedding without compromising antigenicity. Rinehart and Abul-Haj (1951) found that lipid preservation in the tissue was much better with PEG than with paraffin embedding.[19] Antigenic and enzymatic activity was stable in PEG after storage for 6 months.[16,21]

Higher molecular weight PEG with higher melting points will translate to harder blocks and potentially thinner sections; however, it may not be necessary to use higher molecular weights in order to get sections which are thin enough for immunocytochemical localization purposes. Investigators reported that sections in the range of 2 to 10 μm for PEG 1000[1,26] were obtained at room temperature on a rotary microtome, and 2 μm was routine for PEG 600 cryostat sections.[16] The melting points will increase with higher molecular weight PEG and, therefore, higher temperatures and longer infiltration times will be necessary. In some cases, these temperature differences may make a difference in retaining antigenicity and enzymatic activity. In addition, milder fixation protocols can be used with milder, lower temperature embedding methods, which will in turn facilitate the maintenance of antigen reactivity and enzymatic activity. Differing mechanical properties of tissues may dictate changes in the PEG molecular weight and/or fixation and infiltration protocols. The hardness of the block can be varied by mixing different PEG molecular weights[3] and changing the proportions to adjust sectioning quality. In addition, heating or cooling the PEG blocks will change the block hardness and sectioning quality. Although Bard and Ross[1] (1986) reported that PEG was incompatible with tissues that contained high amounts of connective tissues or tissues that harden significantly after fixation, other investigators have successfully used PEG to embed a range of tissues including lung, muscle, arteries,[19] and decalcified bone.[14]

The choice of PEG molecular weight and method will ultimately be based on the mechanical properties of the tissue, the lability of the antigen/enzyme to be localized, and the availability of equipment (cryostat vs. microtome).

A. PEG 1000 AND 1500

In 1972, Mazurkiewicz and Nakane[15] (1972) used PEG 1000 as the embedding matrix to localize growth hormone and luteinizing hormone by

TABLE 1
PEG Used in Immunocytochemistry and Enzyme Histochemistry

PEG mol. wt.	Antigen or enzyme	Tissue	Localization	Ref. method
PEG 600	Cell-CAM 105	Rat liver	Immunoperoxidase	Mowery et al.[16,17]
	DPP IV		LM and EM	
	Other membrane antigens			
PEG 1000	Growth hormone	Rat anterior	Immunoperoxidase	Mazurkewicz and Nakane[15]
	Luteinizing hormone	Pituitary	LM and EM	
	Nerve growth factor	Brain	Immunoperoxidase	Pickel et al.[18]
	Acid phosphatase	Kidney	Enzyme	Manning and Butler[14]
		Pubic symphysis	Histochemistry	
	Neural crest antigens	Chick embryo	Immunoperoxidase ^{125}iodinated Ab	Bard and Ross[1]
PEG 1500	Glucagon	Bovine pancreas	Immunoperoxidase	Bosmon and Go[3]
	Retinol binding protein	Chorionic membrane	Immunogold	Gao and Godkin[8]
	Various antigens	Chick embryo	Immunofluorescence	Thibodeau et al.[24]
PEG 4000	Tubulin actin	Rat testis	Immunofluorescence	Wolosewick[25]

immunoperoxidase in the anterior pituitary gland of the rat for both LM and EM. They identified a major advantage to PEG embedding, which is that tissue sections can be immunolabeled, examined at the light level, and then reembedded for EM. Their methodology was used thereafter by several investigators with some modifications,[23] for immunofluorescence and immunoperoxidase. PEG 1000 was also used for localization of phosphatase activity by enzyme histochemistry in embryonic chicken liver at the LM and EM levels.[20] PEG 1500 was used for the ultrastructural localization of glucagon in pancreatic islets.[3] In addition, mixtures of PEG 1000 and 1500[5] have also been used for embedding.

Bard and Ross[1] (1986) turned to PEG 1000 sections to screen monoclonal antibodies (see Chapter 5, this volume) to differentiated neural crest cells in the chick embryonic eye when frozen sections yielded poor cellular detail which could not provide precise localization of antigens. In addition to better tissue morphology, PEG 1000 sections of 5 to 10 μm yielded 5 times more binding for one of the antigens tested of ^{125}I-labeled antibodies than paraffin, and about 30% more than frozen sections. Thibideau et al.[24] used PEG 1500 for a monoclonal antibody screen against various differentiation specific antigens of the chick embryo.

B. PEG 600

PEG 600 embedding and immunoperoxidase with monoclonal antibodies was used by this author[16,17] to study the localization of plasma membrane antigens of rat hepatocytes. Because lower temperatures could be used for embedding with PEG 600 (melting point, 22°C), dehydration and infiltration steps were performed at room temperature and embedding was accomplished by freezing blocks and sectioning in a cryostat.[9] In that case, the infiltration conditions were mild, and a relatively mild fixation of 2% paraformaldehyde and 0.1% glutaraldehyde, which was thought to be less likely to compromise antigenicity, was successfully used. Sections were easier to cut at 2 to 3 μm than conventional frozen sections. Hematoxylin and eosin staining of a 2 μm PEG section of perfusion fixed rat liver is shown in Figure 1A. Nuclei are intact and very little freezing damage is observed. An antigen which is localized in all membrane domains of the rat hepatocyte plasma membrane is shown in Figure 1B, and one which is confined to the bile canalicular membrane in Figure 1C. The localization of cell-CAM 105 is shown at the ultrastructural level in Figure 2. It is confined to the bile canaliculus. All organelles can be identified and morphological preservation is excellent. Even though mild fixation was used, no diffusion artifacts were seen. Bile canalicular antigens did not diffuse to other membrane domains and lateral antigens (Figure 3) did not diffuse into the bile canalicular domain.

C. PEG 4000 AND ABOVE

Although PEG 4000[26] and PEG 6000[12] have been used to embed tissue for EM observation, they have not been extensively used for immunocytochemistry and histochemistry because of higher temperatures and longer time periods necessary for infiltration. Wolosewick and de Mey[25] (1982) used PEG

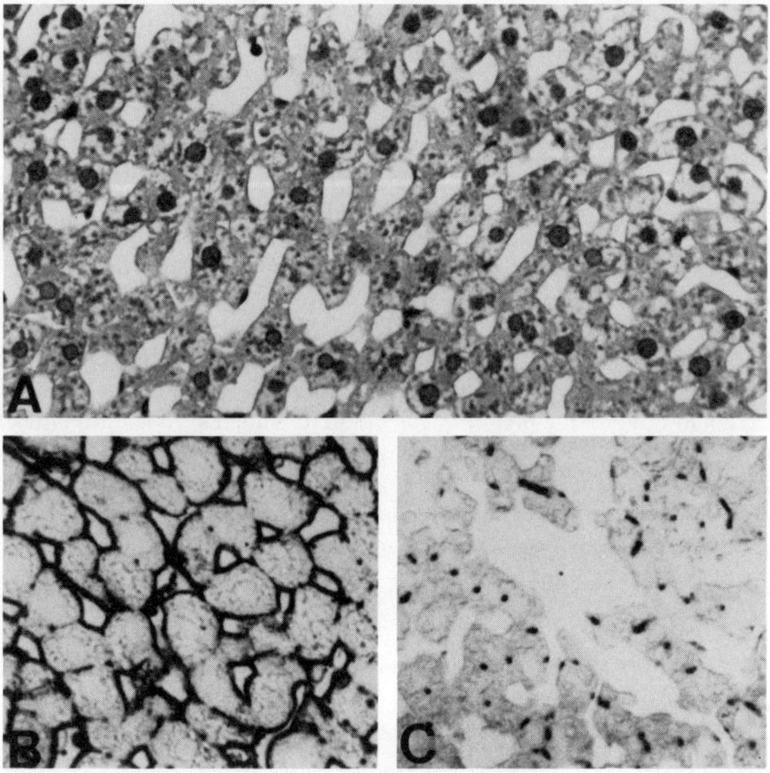

FIGURE 1. LM of 2 μm PEG 600 frozen sections of perfusion fixed rat liver stained with hematoxylin and eosin (A) or labeled by immunoperoxidase as described[16] with MAb 4.3 (B), which is found in all plasma membrane domains of rat hepatocyte, and MAb 5.4 (C) which is found only in bile canalicular domain. Original magnifications: (A) × 250, (B) × 280, (C) × 250. (From Mowery, J., Chesner, J., Spangenberger, S., and Hixson, D. C., Polyethylene glycol as an embedment for microscopy and histochemistry, *J. Histochem. Cytochem.*, 37, 1549, 1989. With permission.)

4000 as a means to cut 0.1 to 0.2 μm sections of rat testis to localize tubulin and actin in the seminiferous epithelium of rat testis by immunofluorescence and then extended the PEG 4000 method to the ultrastructural level by immunogold staining. Mixtures of PEG 4000 and 1000 have also been used by some investigators. Acid phosphatase activity was demonstrated in mouse pubic symphysis and kidney tissue using an embedding mixture of PEG 1000/4000.[14] Although PEG of high molecular weight has been used successfully by these investigators, many antigens do not stand up to the higher temperatures and extended infiltration times needed with higher molecular weight PEGs. In fact, with these higher molecular weights of PEG, dehydration and hardening of tissue can occur during infiltration,[19] and depending on ambient humidity, blocks can be too brittle to section easily.

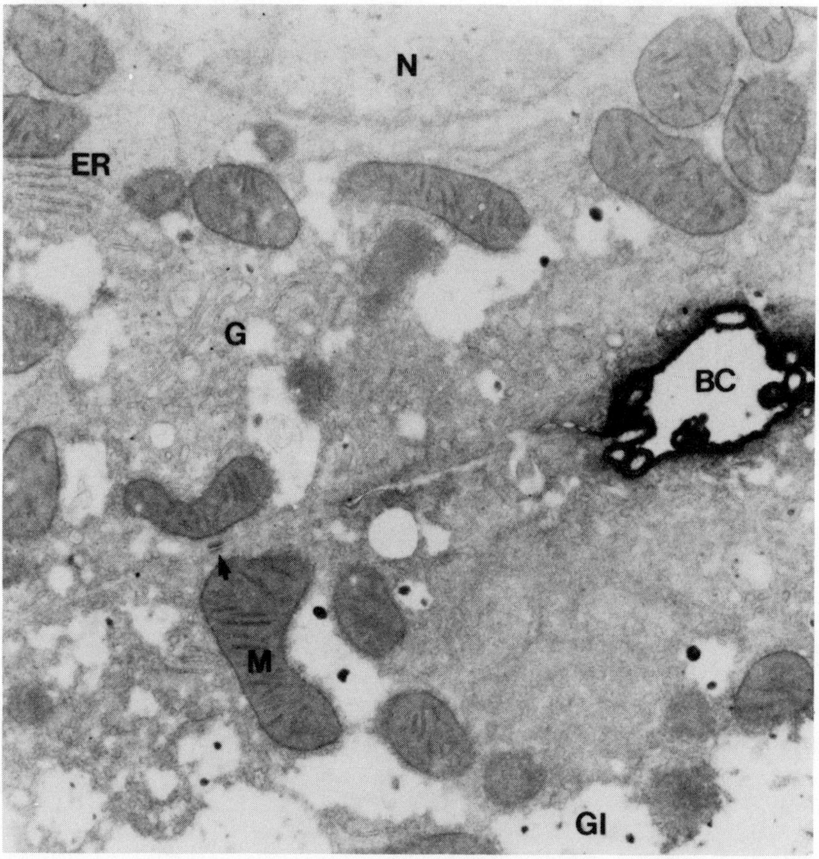

FIGURE 2. Immunoperoxidase localization of cell-CAM 105, bile canalicular membrane protein of rat hepatocyte. Electron micrograph of perfusion fixed rat liver initially embedded in PEG, cryosectioned at 2 μm, immobilized in agarose as described[6,16] and immunoperoxidase labeled with MAb 5.4.[17] Agarose pieces containing sections were then post-fixed in 2.5% glutaraldehyde, osmium tetroxide, flat embedded in Spurr's low viscosity epoxy resin, and sectioned for EM. N, nucleus; ER, endoplasmic reticulum; G, Golgi complex, M, mitochondria, Gl, glycogen; BC, bile canaliculi; arrow, desmosomal junction. (Original magnification × 40,000.) (From Mowery, J., Chesner, J., Spangenberger, S., and Hixson, D. C., Polyethylene glycol as an embedment for microscopy and histochemistry, *J. Histochem. Cytochem.*, 37, 1549, 1989. With permission.)

In short, unless very thin sections are needed, or other PEG methods have been exhausted, the higher molecular weight PEGs are probably not necessary or desirable for most tissues.

III. TECHNICAL PROCEDURES

Most protocols for PEG embedding include primary fixation, dehydration with either PEG or alcohol, infiltration with PEG, and final hardening of the

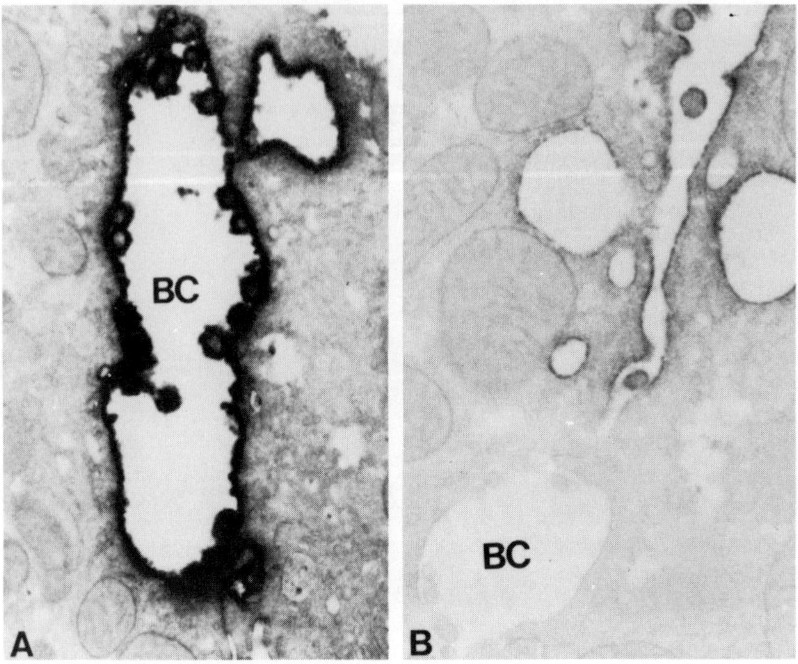

FIGURE 3. Immunoperoxidase localization of domain-specific markers of hepatocyte plasma membrane. Electron micrograph of PEG section labeled as described in Figure 2 but with MAb 4.3(A), which labels all membrane domains and MAb 236.4, which labels lateral and sinusoidal domains. BC, bile canaliculus. [Original magnifications, (A) × 40,000; (B) × 50,000.] (From Mowery, J., Chesner, J., Spangenberger, S., and Hixson, D. C., Polyethylene glycol as an embedment for microscopy and histochemistry, *J. Histochem. Cytochem.*, 37, 1549, 1989. With permission.)

blocks. An outline of three representative protocols for PEG embedding prior to immunocytochemistry is shown in Table 2. Alternative PEG molecular weights will dictate different temperatures for each step. The protocols vary in length, but even the longest ones can be accomplished within 8 h.[16]

A. FIXATION

The method of fixation will obviously have a profound effect on retention of antigenicity and enzyme activity. A detailed discussion of fixation methods is beyond the scope of this chapter, and the reader is referred to the volume by Larsson[13] for an excellent review of characteristics of different fixation methods. There is nearly universal agreement that appropriate fixation must be empirically determined for each antigen/antibody combination. This is especially true for antigenic localizations involving monoclonal antibodies. Because polyclonal antibodies encompass a range of antibodies to different epitopes, not all epitopes will be labile to the same fixative, and antibody concentration can be increased accordingly. Perfusion fixation should be used

TABLE 2
PEG Embedding Procedures

Method A: PEG 1000 (Mazurkiwiez and Nakane[15])

1. Dehydration in graded ethanol series or dehydration in graded series of PEG 1000 in water at 40°C.
2. Infiltration (2 changes) and embedding in 100% PEG 1000.
3. Harden blocks at 5°C.
4. Section on microtome at room temperature.

Method B: PEG 1000 (Bard and Ross[1])

1. Dehydrate in 1:1 water/PEG 1000 at room temperature until tissue sinks (~1 h)
2. Infiltrate with three changes of 100% PEG for 30 min each at 40°C.
3. Embed in fresh PEG at 20°C.
4. Harden blocks further at 4°C if necessary.

Method C: PEG 600 (Mowery et al.[16,17])

1. Dehydrate in increasing concentration of PEG 200 in water at room temperature.
2. Infiltrate in PEG 200/600 with increasing proportion of PEG 600 and with 2 changes of 100% PEG 600 at room temperature.
3. Harden by placed blocks at −20°C.
4. Section in a cryostat.

Method D: PEG 1500 (Thibodeau et al.[24])

1. Dehydration in alcohol.
2. Immerse specimen in PEG 1000 at 45°C for 1 h.
3. Immerse specimen in PEG 1500 at 45°C for 1 h.
4. Embed in fresh PEG 1500 and allow to cool at room temperature to produce blocks.

wherever possible and followed by an additional period of time for immersion fixation. As PEG embedding conditions are relatively mild, milder fixation can be used that would not necessarily stand up to the rigors of harsh embedding conditions.

B. DEHYDRATION

Some procedures do not include a dehydration series and progress immediately to infiltration in 100% PEG;[2,10] however, a dehydration series is generally recommended.[23] The author's prejudice is that a dehydration series in PEG is preferable unless otherwise indicated, although dehydration in alcohol has been used successfully by several investigators.[15,24,26] Mazurkiewicz and Nakane[15] (1972) reported that anterior pituitary glands were easier to section when dehydrated in ethanol.

C. INFILTRATION AND EMBEDDING

The most common method for PEG 1000 embedding is the one recommended by Mazurkiewicz and Nakane[15] (1972) (See Table 2, Method A).

The method (Table 2, Method B) of Bard and Ross[1] (1986) is very similar, but involves a shorter dehydration series. The method of Bosmon and Go[3] (1981) for PEG 1500 involves placing the tissue directly into PEG 1500 with 3 to 5% water. By applying a vacuum oven it is possible to speed the infiltration process for PEG at higher temperatures. However, this is not necessary for lower temperature embedding.

In the PEG 600 embedding procedure in Method C, dehydration and infiltration are accomplished by a series of increasing concentrations of PEG 200, then a switch to a PEG 200:600 mixture, with increase proportions of PEG 600, followed by two changes of 100% PEG 600. Because PEG 600 has a melting point of 22°C and a low viscosity, the entire procedure can be accomplished easily in 8 h at room temperature. For embedding, a small drop of PEG 600 is placed in a BEEM™ capsule and frozen so that when the tissue block is added, it is not resting on the bottom of the capsule. Blocks are stored at −20°C.

D. SECTIONING

The choice of PEG molecular weight will determine the sectioning method used. PEG 1000, PEG 1500, or mixtures of PEG 1000/4000 may be sectioned on a microtome at room temperature. It was reported that sections in the range of 2 to 10 μm were routine for PEG 1000[1,26] at room temperature on a rotary microtome, while 2 μm sections were achieved for PEG 600 cryostat sections.[16] Wolosewick and de Mey used an ultramicrotome to cut 0.1 μm sections with PEG 4000.[26] Some investigators chill the blocks before cutting at room temperature for easier sectioning.

PEG 600 blocks must be stored and sectioned frozen, preferably at −20°C. Liquid PEG 600 is used to attach the block to the cryostat chuck in an analogous manner as O.C.T. (Miles, Kankakee, IL) is used in standard frozen sectioning. With practice, routine frozen sections of 2 to 3 μm can be obtained, which are easier to cut than conventional frozen sections,[16] and 1 μm sections are possible (Mowery, J., unpublished observations).

E. MOUNTING SECTIONS

One of the major challenges for PEG embedding is mounting the sections. Because of its water solubility, which has the important advantage of rendering clearing with alcohols unnecessary, PEG dissolves in buffer instantly, thereby dispersing the sections or pieces of sections. Slides must be manipulated during immunocytochemical staining procedures, and sections must be firmly attached to slides, so several solutions to this problem have been devised.

1. Slide Coating

Coating the slide with poly-L-lysine,[26] dried albumin and glycerin,[15] gelatin,[15] and rubber cement[2] have been tried. Bard and Ross[1] (1986) precoated the slides with 1:1 albumin/glycerin and dried them at 37°C overnight, then mounted the sections in a fresh coat of albumin/glycerin and dried them overnight at 40°C.

If sections are crumpled, as is the case for some tissues and especially for larger sections, a method is needed that will allow the section to be stretched. Bosmon and Go[3] (1981) pressed the PEG sections onto gelatin-chromealum coated glass slides with a thin paintbrush and heated slides on a 45°C hot plate. After the PEG was melted, a drop of water was applied to stretch the section, and slides were dried overnight in an incubator at 37°C.

2. Wire Loop Transfer

Two methods were developed by Mazurkiewicz and Nakane (1972).[15] In the first one, sections were floated over glycerol and then transferred with a wire loop to albumin-coated slides. Excess liquid was removed with filter paper and slides dried at 40°C. Alternatively, sections were floated on a gelatin solution and transferred with a wire loop to slides and dried at 40°C. The slides were then dipped in a 1% gelatin solution and dried. Unfortunately, the gelatin must be partially removed with 0.1% trypsin before labeling, which can affect antigen reactivity.

3. Immobilizing Sections in Agarose

With PEG 600, an alternative method was used by Mowery et al. (1989).[16] Cryostat sections of 2 μm were not mounted on slides, but were embedded in agarose by a modification of the method of De Camilli[6] (1983) and stained in suspension by immunocytochemical reagents. Briefly, ribbons of 2 μm PEG frozen sections were draped in rows in the center of a slide at room temperature, thereby melting the PEG support and temporarily attaching the section. Coverslips were placed at each end of the slide, and a drop of 3% low-gelling temperature agarose (FMC Bioproducts, Rockland, ME) was placed in the center, on the top of the sections. A slightly warmed slide was placed on the top and clamped at each end, making a flattened sandwich of the agarose. After gelation with cooling for 5 min, the agarose was cut into squares of approximately 1 cm^2 with a razor blade and the agarose pieces were placed in buffer for immunolabeling. The PEG matrix is removed in subsequent washing and blocking steps, and the section does not have to be dried or heated above room temperature. This method would not be useful for performing large numbers of different staining protocols at one time (as in a monoclonal antibody screen), but is very useful for reembedding for EM. Sections can be examined at the light microscopic level to detect optimal staining patterns before they are post-fixed in 3% glutaraldehyde and flat-embedded for EM.

4. Silanated Slides

Many investigators have recently turned to silanated glass microscope slides for covalent binding of tissue sections.[5,8] Through an organosilane coupling reagent, amino groups are bound to glass slides by taking advantage of surface hydroxyl groups on the glass slides. The procedure for silanation is given in Reference 25, or silanated slides may be purchased as described.[8] In the method by Gao and Godkin[8] (see Chapter 7, this volume) sections are first overlaid on the surface of agarose blocks, which dissolves the PEG support, and then the agarose blocks are inverted on silanated slides. After air drying, the slides are immersed in phosphate-buffered saline (PBS), and the agarose block detaches, while the sections remain attached. This method apparently allows much faster and easier transfer of sections to slides than other methods.

IV. DISCUSSION

PEG represents a useful alternative to either paraffin embedding or cryosectioning for ICC and EHC. Compared to paraffin embedding, morphological preservation is at least comparable and in some cases better, especially with respect to lipid preservation. Because the PEG is water soluble, clearing with xylene and alcohols prior to immunocytochemical staining is not necessary. Compared to cryosectioning, morphological preservation with PEG is superior and antigenicity can be as good or better,[2] although this will depend on the antigen/fixative combination. PEG embedding has the additional advantage that sections can be reembedded for electron microscopy after immunocytochemical labeling.[15,16] Even the longer PEG embedding protocols are shorter than the lengthy cryoprotective steps that must be performed if frozen sections are to be used for EM observation. Because PEG is relatively soft and hygroscopic, some experimentation is necessary to determine the right protocol and molecular weight to match the hardness of the tissue in question. However, because a variety of tissues have been successfully embedded and sectioned, the investigator can be confident that a workable procedure can be found. Although mounting sections has traditionally been the most challenging aspect of PEG embedding, this problem has been solved in several creative ways, as delineated above, and should no longer represent a significant barrier to using PEG embedding for ICC and EHC.

REFERENCES

1. **Bard, J. B. L. and Ross, A. S. A.,** Improved method for making high-affinity sections of soft tissue embedded in polyethylene glycol (PEG): its use in screening monoclonal antibodies, *J. Histochem. Cytochem.*, 34, 1237, 1986.
2. **Blank, H. and McCarthy, P. L.,** A general method for preparing histologic sections with a water soluble wax, *J. Lab. Clin. Med.*, 36, 776, 1950.
3. **Bosman, F. T. and Go, P. M.,** Polyethylene glycol embedded tissue sections for immunoelectronmicroscopy, *Histochemistry,* 73, 195, 1981.
4. **Carsten, M. E.,** A rapid procedure for the embedding of eyes, *Arch. Pathol.*, 44, 96, 1947.

5. **Clayton, D. F. and Alvarez-Buylla, A.,** *In situ* hybridization using PEG-embedded tissue and riboprobes: Increased cellular detail coupled with high sensitivity, *J. Histochem. Cytochem.,* 37, 389, 1989.
6. **De Camilli, P., Harris, S. M., Huttner, W. B., and Greengard, P.,** Synapsin I (Protein I), a Nerve Terminal-Specific Phosphoprotein II. Its specific association with synaptic vesicles demonstrated by immunocytochemistry in agarose-embedded synaptosomes, *J. Cell Biol.,* 96, 1355, 1983.
7. **Firminger, H. I.,** Carbowax® embedding for obtaining thin tissue sections and study of intracellular lipids, *Stain Technol.,* 25, 121, 1950.
8. **Gao, K. and Godkin, J. D.,** A new method for transfer of polyethylene glycol-embedded tissue sections to silanated slides for immunocytochemistry, *J. Histochem. Cytochem.,* 39, 537, 1991.
9. **Hale, A. J.,** The effect of temperature and of relative humidity on sectioning of tissues embedded in polyethylene glycol wax, *Stain Technol.,* 27, 189, 1952.
10. **Jones, R. M. and Goodale, F.,** Improvements in the method of embedding tissues in Carbowax®, *Am. J. Clin. Pathol.,* 38, 173, 1962.
11. **Kohler, G. and Milstein, C.,** Continuous culture of fused cells secreting antibody of predefined specificity, *Nature,* 256, 495, 1975.
12. **Kondo, H., Wolosewick, J. J., and Pappas, G. D.,** The microtrabecular lattice of the adrenal medulla revealed by polyethylene glycol embedding and stereo electron microscopy, *J. Neurosci.,* 2, 57, 1982.
13. **Larsson, L. I.,** *Fixatives and Tissue Pretreatment in Immunocytochemistry: Theory and Practice,* CRC Press, Boca Raton, FL, 1988, chap. 2.
14. **Manning, J. P. and Butler, M. C.,** Simultaneous formalin fixation and EDTA decalcification with Carbowax® embedding for preservation of acid phosphatase, *Stain Technol.,* 40, 7, 1965.
15. **Mazurkiewicz, J. E. and Nakane, P. K.,** Light electron microscopic localization of antigens in tissues embedded in polyethylene glycol with a peroxidase-labeled antibody method, *J. Histochem. Cytochem.,* 20, 969, 1972.
16. **Mowery, J., Chesner, J., Spangenberger, S., and Hixson, D. C.,** Rapid low molecular weight polyethylene glycol embedding protocol for immunocytochemistry, *J. Histochem. Cytochem.,* 37, 1549, 1989.
17. **Mowery, J. and Hixson, D. C.,** Detection of cell-CAM 105 in the pericanalicular domain of the rat hepatocyte plasma membrane, *Hepatology,* 13, 47, 1991.
18. **Pickel, V. M., Tong, H. J., and Reis, D. J.,** Immunohistochemical localization of tyrosine hydroxylase brain by light and electron microscopy, *Brain Res.,* 85, 295, 1975.
19. **Rinehart, J. F. and Abul-Haj, S.,** Histological demonstration of lipids in tissue after dehydration and embedding in a polyethylene glycol, *Arch. Pathol.,* 51, 666, 1951.
20. **Sandstrom, B. and Westman, J.,** Non-freezing light and electron microscopic enzyme histochemistry by means of polyethylene glycol embedding, *Histochemie,* 19, 181, 1969.
21. **Sandstrom, B.,** Preservation of some histochemically demonstrable enzyme activities in polyethylene glycol-embedded liver tissue during prolonged storage, *Acta Soc. Med. Ups.,* 75, 274, 1970.
22. **Scholer, J. and Armstrong, W. E.,** Aqueous aldehyde (Faglu) histofluorescence for catecholamines in 2 μm sections using polyethylene glycol embedding, *Brain Res. Bull.,* 9, 27, 1982.
23. **Smithson, K. G., MacVicar, B. A., and Hatton, G. I.,** Polyethylene glycol embedding: a technique compatible with immunocytochemistry, enzyme histochemistry, histofluorescence, and intracellular staining, *J. Neurosci. Methods,* 7, 27, 1983.
24. **Thibodeau, A., Duchaine, J., Simard, J., and Vincent, M.,** Localization of molecules with restricted patterns of expression in morphogenesis: an immunohistochemical approach, *Histochem. J.,* 21, 348, 1989.
25. **Wolosewick, J. and de Mey, J.,** Localization of tubulin and actin in polyethylene glycol embedded rat seminiferous epithelium, *Biol. Cell,* 44, 85, 1982.
26. **Uhl, G. R.,** In Situ *Hybridization in Brain,* Plenum Press, New York, 1986, 267.

Chapter 4

PEG EMBEDDING AND SUBSEQUENT DE-EMBEDDING AS A NEW METHOD FOR MULTIPURPOSES IN ELECTRON MICROSCOPY

Hisatake Kondo

TABLE OF CONTENTS

0-8493-4323-2/93/$0.00 + $.50

I. HISTORICAL OVERVIEW

Since epoxy resins were introduced and employed worldwide as an easily obtainable and reasonably reliable crosslinking embedment for transmission electron microscopy (TEM),[17] investigators had to overcome a problem of inadequate contrast of specimens, mainly because cured epoxy resins themselves have a substantial level of electron density. This problem was sufficiently overcome by heavy atom staining of specimens.[19] However, we still have to take into consideration that we may overlook structured elements whose contrast remains similar to, or below that of, the cured epoxy resin even after the thin sections were stained with heavy atoms. The next issue to be considered is the fact that the thin section represents a very small fragment of the cell, and thus a good understanding of the organization in cells requires viewing the depth dimension of the entire living unit. For this issue to be resolved, the viewing of whole mount and unembedded cultured cells has been attempted from the early stage of electron microscopy (EM) history, and it has eventually been made possible by employing high voltage-electron microscopes.[22,23,28]

As a result of this attempt, two rather unexpected findings have been obtained in addition to information on the depth dimension of cells: (1) under embedment-free conditions, any cellular structural elements produce adequate contrast for differential recognition without additional heavy atom staining in EM, and (2) the intracellular compartment, previously termed the cytoplasmic matrix, is occupied with a three-dimensional network of filamentous components, originally termed microtrabecular lattice (Figures 1 through 3). The latter finding raised a serious question whether the filamentous components in the cytoplasm might be artifacts caused by the specimen preparation procedure because of their unexplored structural features. It furthermore revived the early controversy over the structure of the cytoplasm, that is, the structured idea vs. the solution one.[4,20]

For the question of the artifact to be resolved, several attempts have been performed such as examination of cells subjected to various chemical and physical fixation procedures, examination of model systems including erythrocytes and artificial protein solutions after similar fixations, as well as comparison of the images obtained from whole cells with those from conventionally prepared epoxy resin sections of cells. In addition, it was necessary to extend the examinations of the cytoplasmic matrix to various *in situ* cells. This was the reason why an extractable embedment, polyethylene glycol (PEG) was applied as a transient embedding medium to EM,[27] although PEG had already been applied to light microscopy (LM).[1,26] Subsequent EM studies have shown that the PEG-sectioned cells devoid of embedding resin present an image of ultrastructure identical to the image of whole unembedded and critical point-dried or frozen-dried cultured cells observed by high voltage TEM.[3,7,13,14,24] It has also been clarified by comparison of images in the PEG-section with those in epoxy section that the PEG-processing produces no noticeable alterations of cell ultrastructure, and that the microtrabecular or filamentous elements are the structures already contained in conventional epoxy resin sections of cells, but simply masked by the epoxy embedding media[8] (Figure 4). Furthermore, as discussed later in Section II.C., both PEG-embedding method with chemical fixation and the rapid-freezing, deep-etched replica method without chemical fixation, another embedment-free TEM method developed by Heuser,[5] have been shown to present basically the same structures from the cytoplasm when processed properly.[6,10] It is thus clear that the microtrabeculae in the PEG sections correspond well to the thin interconnecting strands in the frozen replica and that all of them are present, but hardly visible in conventional epoxy sections.

Apart from the controversy on the reality vs. artifact of the filamentous structures in the cytoplasm, the PEG method presents several excellent advantages in EM, which will be described in detail in the following sections.

II. PRACTICAL PROCEDURES

A. BASIC PROCEDURE

The chemical fixation of tissues and cells, followed by dehydration with a graded series of ethanol, are performed according to the conventional EM procedure. At the final step of dehydration, vials containing the fixed tissue blocks in absolute ethanol are placed in an oven at 60°C in order to warm the ethanol. The tissue blocks are transferred in the oven into vials containing a 1:1 mixture of PEG 4000 (average molecular weight 3000 to 3700) and absolute ethanol, and they are kept in the vials for 2 to 3 h with vial lids open to let ethanol evaporate gradually. The tissue blocks are then transferred into pure molten PEG contained in gelatin capsules that have already been kept in the oven and well dried. After the blocks sink to the bottom of the capsules, individual capsules are held by long-armed forceps and quickly immersed in liquid nitrogen, resulting in whitish solidification of PEG. The gelatin capsule is removed with a razor blade to expose the solidified PEG. Portions of PEG containing tissue blocks are cut out and mounted with dental wax on supporting stubs fitting the collet of the ultramicrotome. The PEG blocks are trimmed with a razor blade to obtain tissue-faces of 1 to 4 mm, and sectioned with a well dried glass or diamond knife in a room at low humidity. With some practice, long ribbons of wrinkle-free sections can be easily obtained. Sections are picked up with a platinum loop filled with 2.5% sucrose, and put onto glass slides which have been treated with 0.1% poly-L-lysine in distilled water, pH 8.0, for LM, or onto formvar coated, carbon-stabilized grids previously treated with the poly-L-lysine solution for EM.

The thickness of sections can be estimated only by reading units of advance of the microtome. By this estimation, sections two times thicker than the usual ultrathin sections could be made. Usually sections are cut at a setting of 10 to 15 units of advance on the MT-1 ultramicrotome. Immediately after

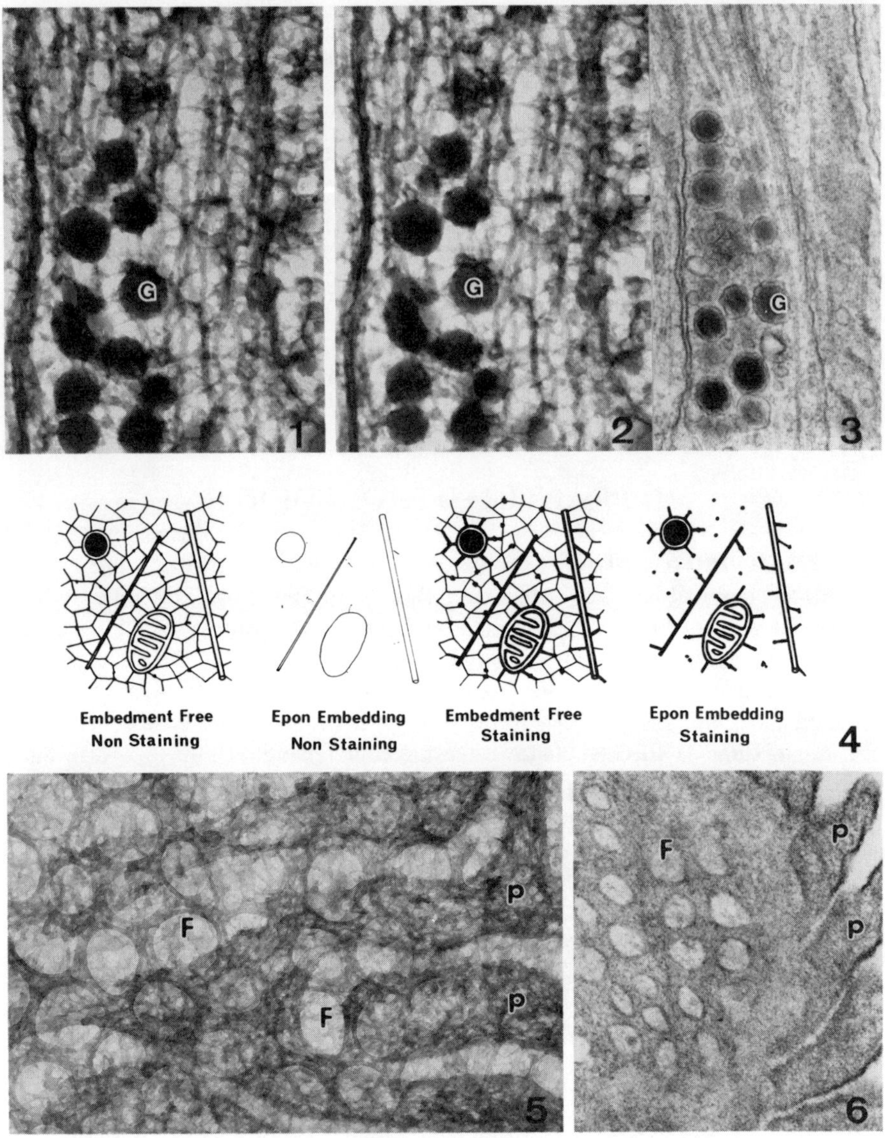

FIGURES 1 through 6.

mounting on grids, the sections on grids are put into a submerged multiple grid holder in 25% ethanol. The grids in the holder are transferred to a graded series of ethanol up to 100%, and subsequently to the critical point-apparatus and dried with liquid CO_2. The dried grids are stored over silica gels until use for EM.

Figures 5 and 6 show well how much the epoxy resin makes the cell structures obscure. While either endothelial cell fenestrae or pedicles of po-

FIGURES 1 and 2. Stereoelectron micrographs of unstained PEG sections of hypophyseal neurosecretory axons. In stereo, neurosecretory granules (G) are interconnected with microtubules by filamentous or microtrabecular strands. (Magnification × 48,300.) **FIGURE 3.** Electron micrograph of stained conventional section of neurosecretory axons embedded in Epon. No clear visibility of interconnecting strands in epoxy section. (Magnification × 28,000.) **FIGURE 4.** Schematic drawings of difference in electron microscopic images between epoxy embedded and embedment-free sections with or without staining. **FIGURE 5.** Electron micrograph of unstained PEG section grazing renal glomerular filtration membrane. Because of increased electron transparency of tissues due to absence of embedding media, superimposed podocyte pedicles (P), basement lamina and endothelial fenestrae (F) are all looked through in this section. Enface view of basement lamina is recognized as forms of meshworks in areas where endothelial fenestrae and interpedicle spaces are superimposed. (Magnification × 48,300.) **FIGURE 6.** Electron micrograph of stained epoxy section of glomerular area comparable to that shown in Figure 5. Because of existence of Epon with inherent electron density, either endothelial fenestrae (F) or podocyte pedicles (P) are visible in area where both of these elements are presumed to be superimposed. (Magnification × 35,700.)

docytes are visible in sections grazing the renal glomerular filtration membrane embedded in epon, it is possible to look through all of the three superimposed elements — the endothelial fenestrae, the basement lamina, and the podocyte pedicles — in grazing embedment-free sections of the filtration membrane. With PEG method the meshwork structure of the basement lamina was clarified for the first time by the present author.[12]

By comparison of the electron micrographs of PEG sections with those of epoxy sections, the difference in size of various cellular and intercellular elements should be noted. In rough estimation, the sizes of bulky elements such as podocyte pedicles and endothelial cells reduce by 30%, whereas the sizes of surrounding empty spaces, such as interpedicel spaces and endothelial fenestrae, increase correspondingly. Such dimensional changes of similar ranges have been noted to occur in specimen preparations for scanning electron microscopy (SEM), and even in the rapid-frozen, deep-etched replicas.[12] Thus, it is likely that the dimensional changes are due to the dehydration procedure and that the etching itself is essentially equivalent to the dehydration. As a result, the dimensional changes under discussion must be unavoidable in the preparation procedure of the embedment-free preparations, regardless of techniques employed.

B. IMMUNOHISTOCHEMISTRY

Although the potential of the PEG method for immunohistochemistry (IHC) at LM level has been recognized previously,[18] the extension of this methodology to EM immunohistochemistry was actually made possible by several authors.[2,9,29]

PEG sections on glass slides pretreated with poly L-lysine are dipped into a 0.1% bovine serum albumin (BSA) buffer (20 m*M* Tris, 0.9% saline, 0.1% BSA, pH 8.2) and processed according to the routine fluorescence IHC. Because PEG sections are very thin as compared with regular cryostat sections and even paraffin sections, resulting fluorescence microscopic images are of

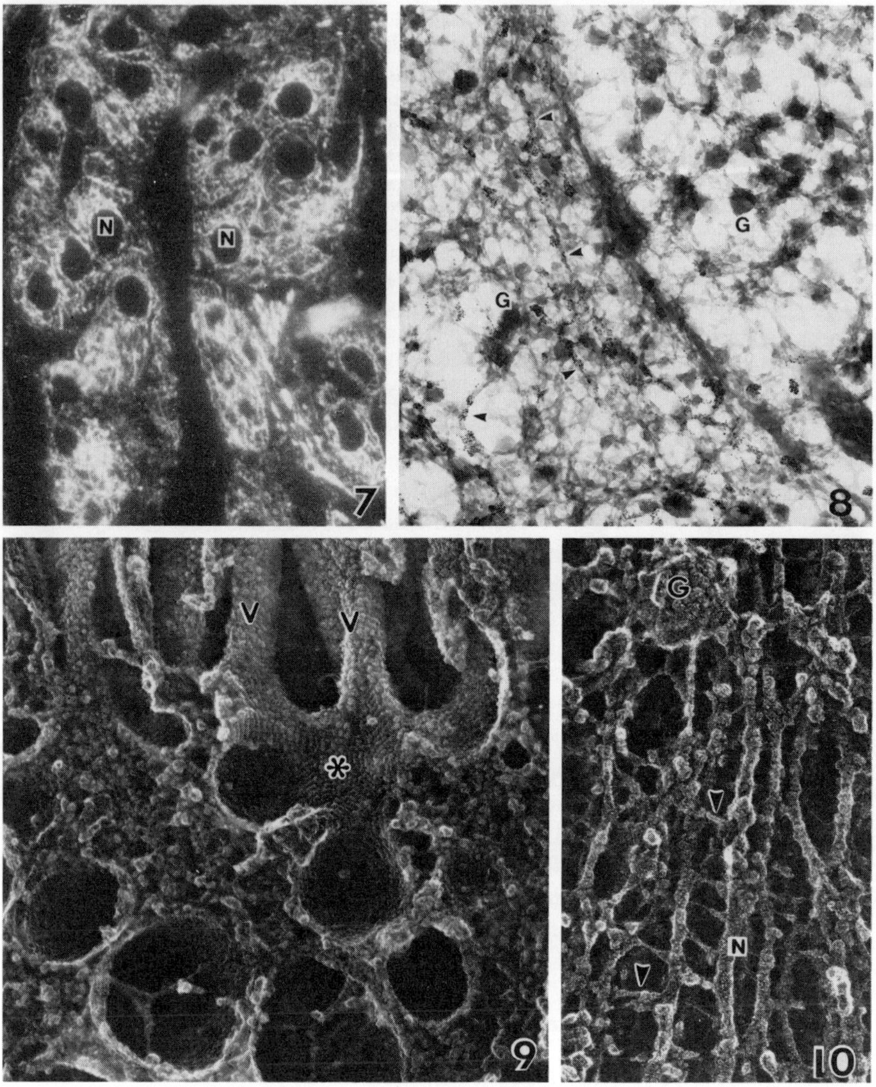

FIGURES 7 through 10.

high quality with excellent resolution, almost comparable to those by confocal laser-scanning microscopy (Figure 7). For immunoelectron microscopy, PEG sections mounted on formvar-coated grids are immersed in 25% ethanol containing sodium borohydride (0.5 mg/ml) for 10 min. After repeated rinse in the BSA buffer and exposure to 1% Triton® X in the same buffer for 30 min, the sections are treated for 20 min with 5% normal goat serum (NGS) in the BSA buffer. The sections are incubated overnight with primary antibodies

FIGURE 7. Immunofluorescence micrograph of PEG section of adrenal medulla immunostained for tubulin. Fine details of individual immunoreactive strands and dots are clearly appreciated in chromaffin cell cytoplasm. (N) chromaffin cell nuclei. (Magnification × 420.) **FIGURE 8.** Immunoelectron micrograph of PEG-sectioned adrenal chromaffin cells immunostained for tubulin. Linear aggregation deposits of gold particles (arrowheads) representing tubulin-immunoreactivity are clearly identified in cytoplasm. (G) chromaffin granules. (Magnification × 28,000.) **FIGURE 9.** Replica electron micrograph of PEG section of intestinal epithelial cell of suckling rat. Distinct geometrical pattern of parallel wavy ridges is evident on cytoplasmic face of endocytic vacuoles (*), and pavements of closely packed granules cover surface of microvilli (V). (Magnification × 48,300.) **FIGURE 10.** Replica electron micrograph of PEG section of hypophyseal neurosecretory axon. Neurotubules (N) and their cross-bridging strands (arrowheads) are visible. (G) neurosecretory granule. (Magnification × 46,200.)

diluted at appropriate concentration in the same buffer containing 1% NGS in a moist chamber at room temperature. After a thorough rinse with BSA buffer, the sections are incubated with IgG-conjugated colloidal gold, 5 nm, diluted at 1:10 in BSA buffer for 3 h in a moist chamber at room temperature. They are repeatedly rinsed with the buffer, post-fixed with 0.5% OsO_4 for 30 s, dehydrated with a graded series of ethanol and then critical point-dried. This approach is quite comparable to, but easier and more reliable than IHC by ultrathin cryosections (Figure 8).

One major drawback of PEG embedding is that heat incubation of specimens in molten PEG may destroy or modify the antigenicity of some soluble proteins, although the antigenicity of most cytoskeletal proteins is preserved well after the PEG method.

C. REPLICA ELECTRON MICROSCOPY

The PEG sections are eventually embedment-free. It is possible to view the surface morphology of every element in the sections by rotary replication with platinum and carbon.

PEG-processed and critical point-dried sections on the formvar-coated grids are transferred into the freeze-fracture apparatus with the specimen-side facing up. The sections are shadowed at −90°C with platinum at an angle of 15° and subsequently are backed with carbon at an angle of 90° while rotating the specimen stage at 60 rpm. The grids are then floated on the surface of household bleach with the specimen-side facing up for 10 min. Following several rinses with distilled water, replicas on the grids are examined under TEM.

Resulting images of the replica in TEM are of high quality and comparable to those prepared through the rapid-freezing, deep-etched replica method (Figures 9 and 10). The similarity in the final images between the two different methods proves this point. This similarity further indicates that the superficial difference in TEM between the PEG section and the frozen replica is due to the fact that the *contour* of every element based on differential contrast is the main information obtained in the section, whereas the *appearance* of the surface morphology of every element is enhanced in the replica. This results in a more fibrous image in the PEG section in comparison to a more granular one in the frozen replica. This superficial difference is well exemplified in Figures 11 and 12. The BSA solution at a concentration of 8% in water presents fibrous networks that are strikingly similar to the cytoplasmic microtrabecular lattice in the PEG section,[6,7] while the same protein solution dominantly takes granular appearance in the PEG replica. From this finding, it seems unlikely that the microtrabecula is an artifact, because the cytoplasmic-soluble proteins are presumed to appear granular. However, this finding does not necessarily lend support to the reality of the microtrabecula, but rather puts serious doubts on its significance and the concept of the structured cytoplasmic matrix. Further consideration on this issue will be found in the articles by Kondo[11,16] and Porter.[21,25]

D. SCANNING ELECTRON MICROSCOPY AND CORRELATION WITH LM AND TEM

By the de-embedding of entire PEG blocks, the PEG method is applicable to SEM, and the intermicroscopic correlation of images is reliably possible among LM, TEM, and SEM.

After taking a section that is used for LM or TEM, the PEG block with its mirror-smooth face is removed from the ultramicrotome and immersed in water. The water is gradually warmed up to 40°C over 10 to 20 min. Subsequently, the de-embedded tissue is refixed with 2.5% glutalaldehyde containing 1% tannic acid in 0.1 *M* Na-cacodylate buffer, pH 7.4, for 1 h, followed by wash in the buffer solution and osmicated with 1% OsO_4 in the same buffer for 5 h. After the postosmication, the tissue blocks are dehydrated in a graded series of ethanol and critical point-dried in CO_2. The dried tissue blocks are mounted on aluminum stubs with silver paste and examined with SEM. The coating with gold palladium and carbon is optional. Figures 13 and 14 show examples of the correlative microscopy. The histological organization is well preserved, the surface morphology of high quality, and the cell interior of one and the same cell are clearly appreciated with this combination (Figures 14 and 15).

III. CONCLUSION

Since its introduction to EM for observation of tissues and cells in embedment-free condition, the PEG method has been extended to IHC at LM and EM levels, to SEM, replica-EM, and the intermicroscopic correlation. Considering the excellent quality of its microscopic images as shown here, PEG method is expected to unravel a new level of histological details as a valuable adjunct to conventional microscopic methodology with continued and wide application, which is summarized in Figure 16.

ACKNOWLEDGMENT

This study was supported by Ueno Clinic Foundation, Tsukuba, Japan.

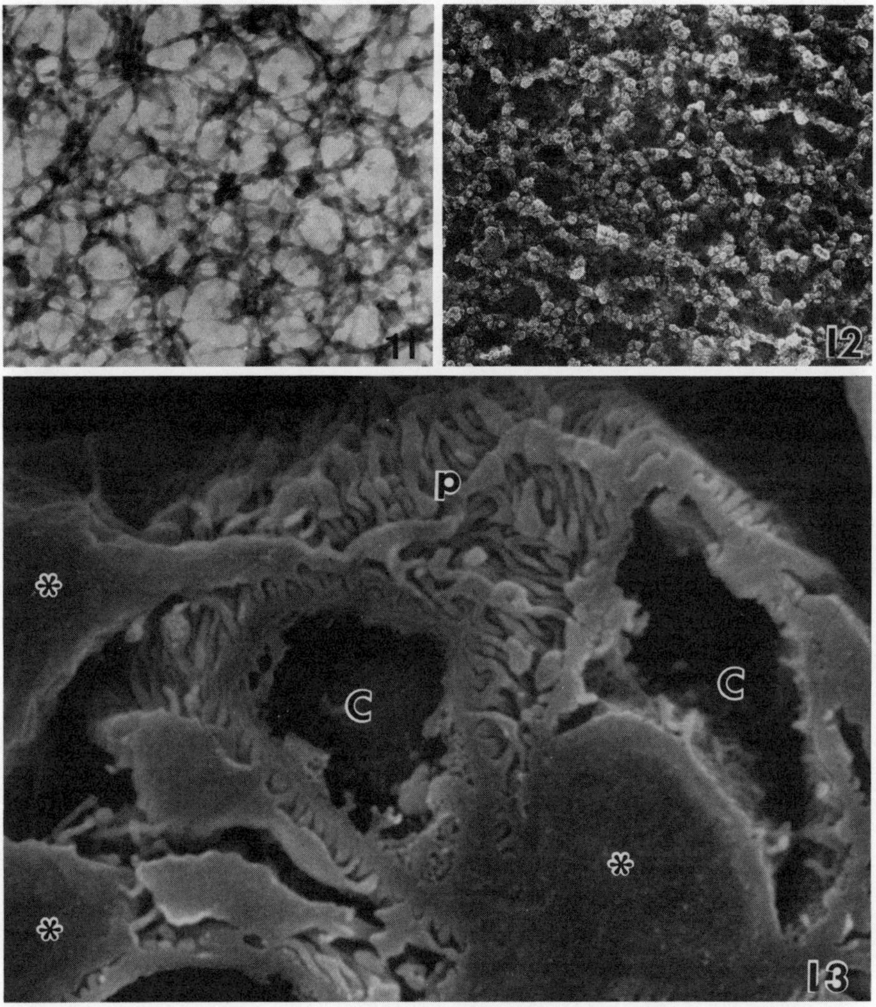

FIGURES 11 through 13.

FIGURE 11. Electron micrograph of PEG section of chemically fixed BSA solution (8%). Fibrous networks similar to intracellular microtrabecular lattices are observed in this embedment-free section. (Magnification × 48,300.) **FIGURE 12.** Replica electron micrograph of PEG section shown in Figure 11. Note enhancement of granular impression of electron microscopic image by replication in contrast to fibrous appearance of chemically fixed protein solution in PEG section. (Magnification × 48,300.) **FIGURE 13.** Scanning electron micrograph of PEG-processed renal glomerulus. Note excellent preservation of surface morphology and structural organization of podocytes (P) and capillaries (C). Cutting face of tissue block after taking out thin sections is marked by asterisks. (Magnification × 7,700.)

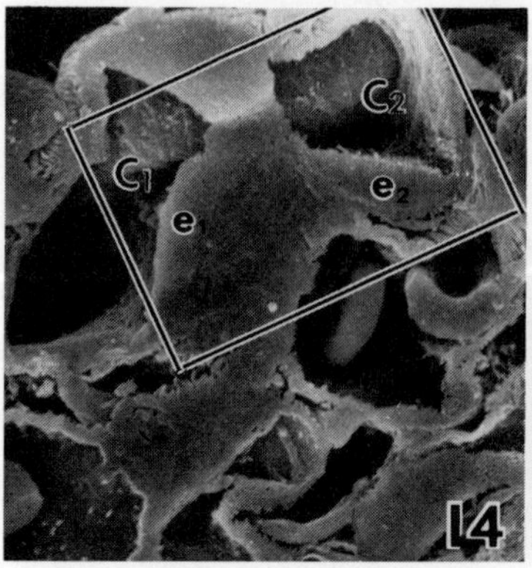

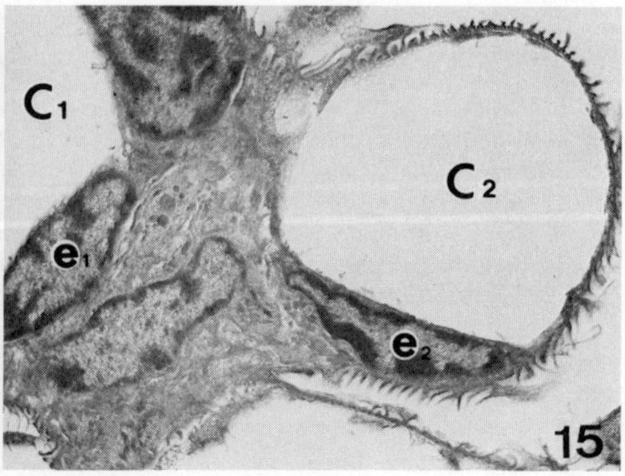

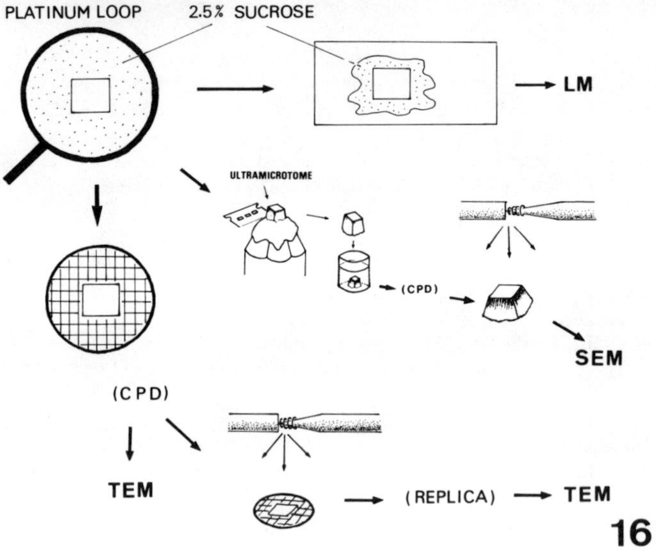

FIGURES 14 through 16.

FIGURE 14. Scanning electron micrograph at low magnification of PEG-processed renal glomerulus. (Magnification × 2,400.) **FIGURE 15.** Electron micrograph of PEG section of area marked by rectangle in Figure 14. This section and tissue-cutting face marked by asterisks in Figure 14 are next to each other, and thus exact correlation of capillaries (C_1, C_2) and endothelium (e_1, e_2) between scanning and transmission electron microscopy is appreciated. (Magnification × 3,400.) **FIGURE 16.** Schematic drawing of usefulness of PEG method for multipurposes in light and electron microscopy. (CPD) critical point drying; (LM) light microscopy; (SEM) scanning electron microscopy; (TEM) transmission electron microscopy.

REFERENCES

1. **Blank, H. and McCarthy, P. L.,** A general method for preparing histologic sections with a water-soluble wax, *J. Lab. Clin. Med.,* 36, 776, 1950.
2. **Bosman, F. T. and Go, P. M.,** Polyethylene glycol embedded tissue sections for immunoelectron microscopy, *Histochemistry,* 73, 195, 1981.
3. **Fontino, M. and Giddings, T. H.,** Ultrastructural visualization of unfixed and unstained whole mounts by high-voltage electron microscopy at low temperature, *J. Ultrastruct. Res.,* 91, 112, 1985.
4. **Hardy, W. B.,** On the structure of cell protoplasm, *J. Physiol.,* 24, 158, 1899.
5. **Heuser, J.,** Three-dimensional visualization of coated vesicle formation in fibroblasts, *J. Cell Biol.,* 84, 560, 1980.
6. **Kondo, H.,** Reexamination of the reality or artifacts of the microtrabeculae, *J. Ultrastruct. Res.,* 87, 124, 1984.
7. **Kondo, H.,** Polyethylene glycol (PEG) embedding and subsequent de-embedding as a method for the structural and immunocytochemical examination of biological specimens by electron microscopy, *J. Electron Microsc. Tech.,* 1, 227, 1984.
8. **Kondo, H.,** What is the microtrabecula? *J. Electron Microsc.,* 34, 123, 1985.
9. **Kondo, H.,** Ultrastructural localization of actin in the intermediate lobe of rat hypophysis, *Biol. Cell,* 60, 57, 1987.
10. **Kondo, H.,** Visualization of the inner and outer surfaces of the cell membrane and cytoskeleton by polyethylene glycol embedding, subsequent deembedding, and rotary replication with platinum, *J. Electron Microsc. Tech.,* 7, 17, 1987.
11. **Kondo, H.,** The cytoplasmic matrix of the adrenal chromaffin cells of rats under normal and stressed conditions, *J. Electron Microsc. Tech.,* 12, 356, 1989.
12. **Kondo, H.,** Rat kidney glomerular basement membrane visualized *in situ* by embedment-free sectioning and subsequent platinum-carbon replication, *J. Electron Microsc. Tech.,* 14, 63, 1990.
13. **Kondo, H., Wolosewick, J. J., and Pappas, G. D.,** The microtrabecular lattice of adrenal medulla revealed by polyethylene glycol embedding and stereo-electron microscopy, *J. Neurosc.,* 2, 57, 1982.
14. **Kondo, H., Pappas, G. D., and Wolosewick, J. J.,** The cytoskeletal lattice of the neurohypophysial cells, *Biol. Cell,* 49, 99, 1983.
15. **Kondo, H. and Ushiki, T.,** Polyethylene glycol (PEG) embedding and subsequent de-embedding as a method for the correlation of light microscopy, scanning electron microscopy, and transmission electron microscopy, *J. Electron Microsc. Tech.,* 2, 457, 1985.
16. **Kondo, H., Yamamoto, M., and Watanabe, M.,** Reversible intracellular displacement of the cytoskeleton protein and organelles by ultracentrifugation of the sympathetic ganglion, *J. Submicrosc. Cytol. Pathol.,* 24, 241, 1992.
17. **Luft, J. H.,** Improvements in epoxy resin embedding methods, *J. Biophys. Biochem. Cytol.,* 9, 409, 1961.
18. **Mazurkiewicz, J. E. and Nakane, P. K.,** Light and electron microscopic localization of antigens embedded in polyethylene glycol with a peroxidase-labeled antibody method, *J. Histoch. Cytochem.,* 20, 969, 1972.
19. **Pease, D. C. and Porter, K. R.,** Electron microscopy and ultramicrotomy, *J. Cell Biol.,* 91, 287s, 1981.
20. **Porter, K. R.,** The cytomatrix; a short history of its study, *J. Cell Biol.,* 99, 3s, 1984.
21. **Porter, K. R.,** Structural organization of the cytomatrix, in *The Organization of Cell Metabolism,* Welch, G. R. and Clegg, J. S., Eds., Plenum Press, New York, 1987.
22. **Porter, K. R., Claude, A., and Fullam, E. F.,** A study of tissue cells by electron microscopy. Methods and preliminary observations, *J. Exp. Med.,* 81, 233, 1945.
23. **Porter, K. R. and Thompson, H. P.,** Some morphological features of cultured rat sarcoma cells as revealed by the electron microscope, *Cancer Res.,* 7, 431, 1947.

24. **Porter, K. R. and Anderson, K. L.,** The structure of the cytoplasmic matrix preserved by freeze-drying and freeze-substitution, *Eur. J. Cell Biol.*, 29, 83, 1982.
25. **Porter, K. R. and McNiven, M. A.,** The cytoplast: a unit structure in chromatophores, *Cell,* 29, 23, 1982.
26. **Riopel, J. L. and Spurr, A. R.,** Carbowax® for embedding and serial sectioning of botanical material, *Stain Technol.*, 37, 357, 1962.
27. **Wolosewick, J. J.,** The application of polyethylene glycol (PEG) to electron microscopy, *J. Cell Biol.*, 86, 675, 1980.
28. **Wolosewick, J. J. and Porter, K. R.,** Microtrabecular lattice of the cytoplasmic ground substance. Artifact or reality, *J. Cell Biol.*, 82, 114, 1979.
29. **Wolosewick, J. J., deMey, J., and Meininger, V.,** Ultrastructural localization of tubulin and actin in polyethylene glycol-embedded rat seminiferous epithelium by immunogold staining, *Biol. Cell,* 49, 219, 1983.

Chapter 5

USING PEG SECTIONS TO SCREEN MONOCLONAL ANTIBODIES

Jonathan B. L. Bard and Allyson S. A. Ross

TABLE OF CONTENTS

0-8493-4323-2/93/$0.00 + $.50

I. INTRODUCTION

Normal paraffin histology works remarkably well. Not only is morphology well preserved, but so are both protein confirmation as assayed by immunohistochemistry (IHC) and DNA and RNA sequences as tested by *in situ* hybridization. These advantages of traditional paraffin histology are so obvious and the technology so routinely established that it is easy to forget the limitations of the technique and that preservation is not actually as impressive as it first seems. The morphology of tissues from paraffin-wax sections is not particularly good as compared to plastic sections, while the treatment of the proteins in the tissue is surprisingly vicious: they are either crosslinked or condensed and then dehydrated and heated to temperatures that would normally denature them. It is not, therefore, surprising that many antibodies do not work particularly well on sections of paraffin-embedded material and frozen sections with their poor morphology sometimes have to be used.

The long-established[3] method of embedding in water-soluble polyethylene glycol (PEG) gives us the opportunity of circumventing some of the problems associated with paraffin-section and frozen-section histology. The quality of tissue preservation is sufficiently good for the sections to be used for transmission electron microscopy (TEM)[9], while, because the tissue need neither be dehydrated nor overheated, protein structure should be well preserved. PEG-embedded material would thus be expected to give excellent results for IHC.[7] Unfortunately, PEG-embedded material is not user-friendly: the water-soluble sections that are difficult to float out are hard to handle (particularly on humid days) and are difficult to affix to slides. Their use has therefore been somewhat limited and until recently restricted to small (~1 mm) thick pieces of tissue. This chapter shows how at least some of these problems can be overcome.[2]

A. THE PROBLEMS TO BE SOLVED

Our use of PEG arose as a result of our setting out to make monoclonal antibodies against antigens on the neural crest cells that invade the developing chick cornea.[5] This meant that we had no choice but to test our antibodies against sectioned material, and we were therefore (and, as it turned out, rightly) concerned that the antigens in the material might alter during fixation or processing and then might no longer bind the antibodies.

The problem that we had to solve was to make large numbers of good-sized sections on a routine basis where the antigens were in a state far closer to that *in vivo* than is usual in paraffin sections whose preparation involves long fixation, considerable temperatures, and complete dehydration. We therefore set out to use minimal fixation, to keep the temperature of the material below 40°C, and to ensure that the material was not dehydrated by solvents. We were ineluctably led to the use of PEG, as they are water soluble and, if their molecular weight is low, will melt at low temperatures.

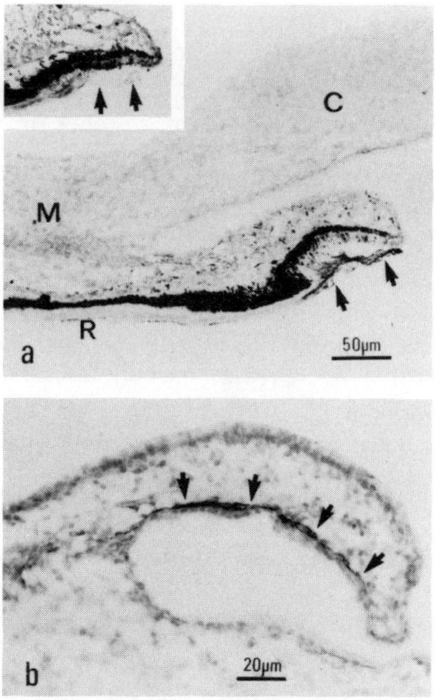

FIGURE 1. Examples of PEG immunohistochemistry. (a) PEG section of peripheral region of eye showing cornea (C), retina (R) and neural-crest-derived mesenchyme (M) at side of eye. Arrows point to region where antibody MA1.84 stains extracellular matrix at tip of retina. Insert shows tip of retina in similar paraffin section; note that immunostaining is far less intense than in PEG section. (Magnification × 140.) (b) PEG section of developing eyelid of day 9 chick embryo stained with MA1.268.2, monoclonal antibody that binds to component of basal lamina (arrow) where rudiment separates from main epithelium. Note that hydrated extracellular matrix in eyelid has not condensed so that tissue maintains hydrated appearance (hollow arrow). (Magnification × 300.) (From Bard, J. B. L. and Ross, A. S. A., Improved method for making high affinity sections of soft tissue embedded in polyethylene glycol (PEG); its use in screening monoclonal antibodies, *J. Histochem. Cytochem.*, 34, 1237, 1986. With permission.)

In this chapter, we discuss the procedures that we used and how we set out to optimize their use so that it was practical to cut large numbers of sections and test 100 or more supernatants a day. We then consider how much better PEG sections were than normal paraffin ones and end by listing the advantages and disadvantages of PEG embedding for testing antibodies.

II. TECHNICAL PROCEDURES

The material that was used for eliciting antibodies in mice was the mesenchyme surrounding the cornea of the day six chick embryo (see Figure 1), as this contains neural-crest-derived cells that will later differentiate into stromal fibroblasts, limbal mesenchyme, and the frontal bones of the head.[5] This mesenchyme was dissected from the embryos and inserted subcutaneously in Swiss mice with Freund's adjuvant in the usual way, and, in due course, successful fusions were made and clones grown (for methods, see Reference 8). Over the next few weeks, we had to test up to 100 supernatants a day, and therefore needed the equivalent number of sections. To facilitate these assays, we used Teflon®-coated multi-well slides (C. A. Hendley, U.K.) in which there were 12 circular spaces each about 1 cm across. Ten of the spaces were used for sections and the other two for labeling.

A. FIXATION

The antibody-containing supernatants had to be tested on sections of the original tissue, as there is no appropriate cell line or ELISA test for these antigens. Heads were removed from 5 to 15-day chick embryos, bisected, and trimmed to leave only the eyes and surrounding mesenchyme. For the reasons previously given, we chose to use only light fixation, the tissues therefore, were fixed in Millory's buffered formalin at room temperature for only 30 min and were then well-washed in phosphate-buffered saline (PBS). The posterior parts of the eye, the vitreous, and the lens were then dissected away because low-temperature PEG embedding is only suitable for soft tissue (see later sections). It should also be said that this short fixation time had an additional benefit: the tissue remained relatively soft and could therefore be supported by low melting point PEG.

B. EMBEDDING

There is a choice of PEGs that can be used for embedding, with molecular weights from 200 to 10,000 and melting points ranging from 20°C to unrealistically high temperatures (>60°C). A balance has to be struck here, as the lower the melting point of PEG used, the softer the block, and the harder it is to cut sections at room temperature. While soft PEG can be frozen and then cut in a cryostat,[7] this procedure is not convenient for cutting large numbers of sections. We therefore compromised by using PEG with a molecular weight of 1000, which has a melting point of 39°C and does not require that the tissue be taken to a temperature higher than it would have encountered *in vivo*. Fortunately, this PEG turns out to be firm enough to support embryonic material, which lacks substantial connective tissue and tough membranes, provided that it has not been too heavily fixed.

The procedure that we adopted was as follows: fixed tissues were first washed in water (3 × 5 min) and left in a 1:1 water/PEG solution until they sank (~1h). They were then given three 30-min changes in PEG 1000 at 40°C and allowed to cool in fresh PEG 1000 at room temperature to produce blocks. These were then cooled to 4°C, at which temperature they were stored in sealed plastic bags until needed. (It seems that higher-melting-point PEG blocks can be stored after dipping them briefly in molten paraffin.)

C. SECTIONING

The problem with cutting PEG blocks is not only that they are soft, but that sections tend to crumple in a humid atmosphere. We therefore found it important that sections were cut on a cool, dry day or in an air-conditioned room, and that the knife was cooled with an ice bag. [An alternative means of cooling the knife that we have not used, but that should work, is to mount a funnel containing dry ice (solid CO_2) above the knife and to use the cold air dropping through the funnel to cool the system.] With these precautions, we found that ribbons of 5 to 10 μm sections could be cut from blocks up to about 8 × 3 mm in cross section, provided that the sections were cut slowly (rapid cutting can melt the sections).

D. MOUNTING

There are two problems associated with mounting PEG sections. As it is soluble in water, PEG sections are not possible to float out in the usual way. Because PEG does not adhere to glass particularly well; it is also hard to ensure that the sections will remain stuck to the slides long enough for the various procedures required for IHC to be carried out. Although adhesivity can be improved with the use of poly-L-lysine[10] or 3-aminopropyl triethoxysilane (TESPA)-coated slides,[4] these "glues" do not allow sections to stretch out and are therefore best used for relatively small sections (~1 mm). We chose to use a more fluid mountant to facilitate the spreading of the section in the following way: sections were coated with a 1:1 mix of albumin and glycerine which was allowed to dry at 37°C overnight. A second coat of the mix was then spread on the slide and the sections mounted in it. The slides were then briefly warmed to 40°C, at which temperature the sections spread out before the second coat dried down. At this stage, the sections could be manipulated with a fine needle to center them and to flatten out any folds (the light fixation meant that the sections were still flexible).

For establishing protocols, ribbons of sections were mounted on normal, albumen/glycerine-coated slides; for testing antibodies in large numbers, however, individual sections were cut from the original ribbon and one or two placed in each of the wells of a multi-spot slide which had been given the double coat of albumen and glycerine. While this mountant had the disadvantage that it could produce background staining unless well washed, we found that it held the sections well enough for all routine procedures.

E. IMMUNOHISTOCHEMISTRY

Slides were first washed in PBS (2 × 5 min) to remove the PEG and then treated with methanol and 0.5% hydrogen peroxide to block any endogenous peroxidase activity. They were then routinely exposed to antibody-containing supernatants, given standard washes, and incubated with peroxidase-labeled sheep anti-mouse immunoglobulin G (IgG)$F(ab')_2$, the peroxidase activity being exposed with 3,3′-diamino-benzidene-4HCl (DAB) and hydrogen peroxide. The sections were then counterstained with 0.1% methyl green, dehydrated with alcohol, dipped in xylene, and air-dried, mounted in DPX® (B.D.H., U.K.), and a coverslip then placed over them. The sections and their stains are stable. The green stain provided a good background color that not only allowed the brown peroxide stain to stand out clearly under the microscope, but produced good black and white micrographs, as well as color ones. Indeed, methyl green stained PEG sections better than paraffin sections, where it tended to give a blue tone and some metachromasia against which it was not always easy to see a peroxide stain.

III. RESULTS

A. MORPHOLOGY

Although we found that some larger sections buckled or folded, we were generally pleased with the morphological quality of our material. In routinely stained sections, not only was there excellent preservation of tissue, but our subjective opinion was that the quality of that detail was better than in paraffin sections. Early embryological material is extremely soft and loose with large volumes of extracellular matrix. As this matrix condenses on solvent dehydration,[1] tissue morphology is often distorted in paraffin-wax blocks. The fact that PEG-embedded tissue is not dehydrated seemed to maintain the general morphology of the head mesenchyme rather better than that of paraffin sections (Figure 1). This effect may have been enhanced by the PEG and the tissue both being fairly soft.

This quality of tissue maintenance was particularly apparent in the retina where the morphology of the differentiating rods in the inner layer was excellent in PEG sections, but disrupted in paraffin sections (Figure 2). Similarly, we found that corneal stromas which are highly hydrated looked far less shrunken in PEG sections than in paraffin sections. We suspect that insufficient attention has been paid to the advantages of PEG embedding for examining such hydrated structures.

B. SENSITIVITY

Antibody staining on PEG sections was usually sharp and well defined. There was thus little difficulty in choosing supernatants that gave results meriting more detailed examination. We were consequently able to identify antibodies that stained retinal cells, sclera, basal lamina, mesenchyme, and other tissues. In short, the sections behaved well.

To confirm the quality of the antibody binding, we also tried specific antibodies on paraffin sections and frozen sections of anterior head tissue, with the latter being of tissue mounted in O.C.T.® (Tissue-Tec, U.S.A.), but in neither case were the results as satisfactory (Figure 1). Frozen sections (which were not particularly easy to cut and were about twice the thickness of the PEG sections) had poor morphology, and, although there was usually clear evidence of staining, it tended to be more diffuse than on PEG sections. In contrast, the general rule was that antibody staining was considerably weaker on paraffin sections than on PEG sections.

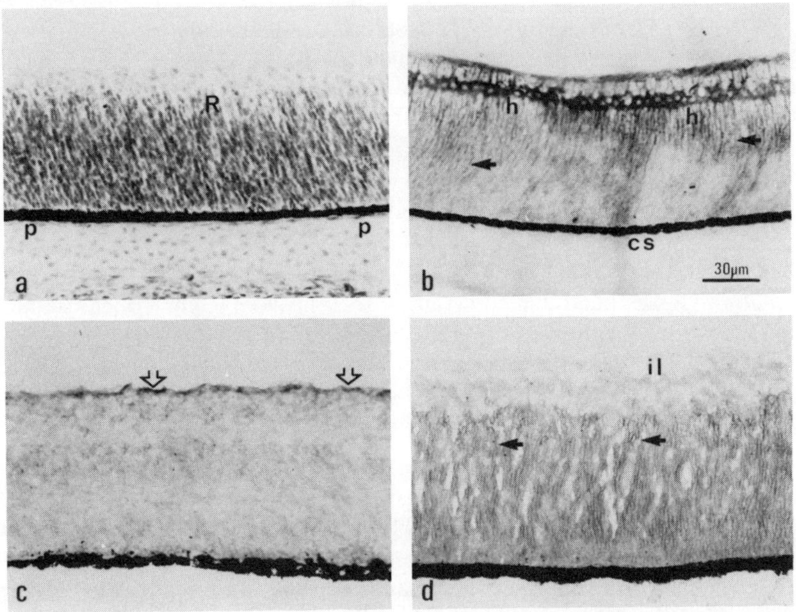

FIGURE 2. Sections of 9-d chick embryonic eye showing neural retina (R) and pigmented retinal epithelium. (a) PEG section stained with methyl green and photographed with red filter shows essential structure of retina. PEG (b), frozen (c), and paraffin (d) sections stained with MA1.189 and photographed with pale green filter. Only in PEG section can either detailed histological structure or clear pattern of antibody staining be seen, together with layer of close adhesion (hollow arrow). Note here excess counter-staining (cs) which often occurs in PEG sections and that is due to mountant taking up small amounts of stain. In frozen section, staining is smeared (even though specimen was cryoprotected with sucrose) and, although bipolar cells (arrow) can be seen, layer of close adhesion has not been preserved. Matters are even worse in paraffin section where staining is weak and inner layer (il) of retina has been disrupted. (All four photos magnification × 360.) (From Bard, J. B. L. and Ross, A. S. A., Improved method for making high affinity sections of soft tissue embedded in polyethylene glycol (PEG); its use in screening monoclonal antibodies, *J. Histochem. Cytochem.*, 34, 1237, 1986. With permission.)

In an attempt to quantify the extent to which protein affinities varied among the different types of sections, we set up an experiment using radio-labeled antibodies which were then allowed to bind to the sections in the usual way; excess antibody was then washed away and the adherent radio-activity counted. For this experiment, we used the serum from a mouse immunized against periocular mesenchyme for the primary antibody and a ^{125}I-labeled rabbit-anti-mouse IgG secondary antibody and counted the adherent radioactivity on a gamma counter.[6]

For a comparison, we took eight paraffin, frozen, and PEG sections (the last with an albumen control) and immunostained them in the usual way, cut the areas containing the sections from the glass slides, and counted the

TABLE 1
The Binding of ^{125}I-Anti-Mouse Antibody to Sections Labeled with Polyclonal Mouse Anti-Chick Antibody

Section type	Number of sections	Total counts	Net counts per section
Paraffin	16	2.3×10^4	1.46×10^3
Frozen	16	8.6×10^4	5.36×10^3
Albumen control	—	4.5×10^4	—
PEG	16	15.4×10^4	7.06×10^3

Note: In this experiment, there were eight sections per slide. Counts of the albumen control were based on two albumen-covered slides each cut to a size equivalent to that of the PEG sections.

associated radioactivity. The results of a typical experiment are shown in Table 1. They show that, although there is little difference in the affinities of the antibodies to PEG and frozen sections (although the former are marginally greater even though the sections are half the thickness), the PEG sections bind about five times as much radioactivity as the paraffin sections, a figure that subjectively confirms the impression from peroxidase-labeled slides. It should be noted that there was some adhesivity of the antibodies to the albumen/glycerine-coated slides, and the amount had to be subtracted from the figure for the PEG sections.

IV. CONCLUSIONS

Our work shows that, although PEG sections are more difficult to cut and manipulate than paraffin or frozen sections, there are distinct advantages in using the PEG embedding method. Their preservation of both morphology and protein structure renders them particularly appropriate for soft embryological tissue. In addition to the technical difficulties, however, there is one problem associated with PEG histology that we have not yet touched on. As tissue is fixed, it hardens, and tissue with crosslinked connective tissue and substantial membranes become particularly hard. One of the purposes of an embedment is to support the tissue as it is cut, and we have found that PEG 1000 is too soft to support ordinary tissues, even when cooled on ice. The solution here is either to freeze the blocks or to use PEG of a higher molecular weight and hence, a higher melting point. In the latter case, work will have to be done to investigate the extent to which antibody affinities drop with increasing embedding temperature.

There is one further use for PEG sections that does not yet seem to have attracted the attention that it merits. With the increasing interest in development that has come with the advent of the techniques of molecular genetics, there is a need for high quality sections that can be used both for IHC and for *in situ* hybridization. As early embryonic tissue is usually small and delicate, it lends itself to PEG embedding, and it will be interesting to see how well the sections stand up to the robust treatment that current *in situ* hybridization techniques demand. Here, PEG sections may be particularly appropriate for nonisotopic RNA hybridization, which combines riboprobe annealing to messenger RNA (mRNA) followed by IHC on the same section.

V. SUMMARY

Although the benefits of PEG embedding are well known and include the high quality of tissue preservation and the maintenance of antigen integrity, the material is hard to use for large pieces of tissues, while standard sectioning techniques do not lend themselves for cutting the large numbers of sections necessary for monoclonal antibody screens. In this chapter, we describe a method of embedding pieces of embryonic material up to about 1 mm thick in PEG and cutting sufficient numbers of sections on a routine basis to test monoclonal antibody supernatants. In particular, we give methods for cutting and mounting the sections and show that, although tissue morphology is as good as or better than that in paraffin sections, the sensitivity of the antigens is about five times as high as for paraffin sections and marginally better than those in frozen sections. We also point to the advantages of PEG sections for maintaining tissue morphology in hydrated tissue and to its potential for *in situ* hybridization studies.

REFERENCES

1. **Bard, J. B. L., McBride, W. H., and Ross, A.,** The morphology of hyaluronidase-sensitive cell coats as seen in the SEM after freeze-drying, *J. Cell Sci.*, 62, 371, 1983.
2. **Bard, J. B. L. and Ross, A. S. A.,** Improved method for making high-affinity sections of soft tissue embedded in polyethylene glycol (PEG): its use in screening monoclonal antibodies, *J. Histochem. Cytochem.*, 34, 1237, 1986.
3. **Blank, H. and McCarthy, P. L.,** A general method for preparing histological sections with a water-soluble wax, *J. Lab. Clin. Med.*, 36, 776, 1950.
4. **Gao, K. and Godkin, J. D.,** A new method for transfer of polyethylene glycol-embedded tissue sections for silanated slides for immunocytochemistry, *J. Histochem. Cytochem.*, 39, 537, 1991.
5. **Johnston, M. C., Noden, D. M., Hazelton, R. D., Coulombre, J. L., and Coulombre, A. J.,** Origins of avian ocular and periocular tissue, *Exp. Eye Res.*, 29, 27, 1979.
6. **Morris, R. J. and Raisman, G.,** Estimation of thy-1 in cryostat sections of nervous tissue, *J. Neurochem.*, 40, 637, 1983.
7. **Mowery, J., Chesner, J., Spangenberger, S., and Hixson, D. C.,** Rapid low molecular weight polyethylene glycol embedding protocol for immunocytochemistry, *J. Histochem. Cytochem.*, 37, 1549, 1989.

8. **Van Heyningen, V., Barron, L., Brock, D. J. H., Crichton, D., and Lawrie, S.,** Monoclonal antibodies to human α-foetoprotein: analysis of the behaviour of three different antibodies, *J. Immunol. Methods,* 50, 123, 1982.
9. **Wolosewick, J. J.,** The application of polyethylene glycol to electron microscopy, *J. Cell Biol.,* 86, 675, 1980.
10. **Wolosewick, J. J. and de May, J.,** Localization of tubulin and actin in polyethylene glycol embedded rat seminiferous epithelium, *Biol. Cell.,* 44, 85, 1982.

Chapter 6

APPLICATION OF PEG SECTIONS FOR *IN SITU* HYBRIDIZATION STUDIES

Arturo Alvarez-Buylla and David F. Clayton

TABLE OF CONTENTS

0-8493-4323-2/93/$0.00 + $.50

I. INTRODUCTION

Polyethylene glycol (PEG) is a water soluble wax with unique advantages for embedding biological materials. Many of these advantages are discussed in other chapters of this book. *In situ* hybridization allows the localization of specific RNA molecules in tissue sections. With this technique, the presence of specific messenger (mRNA) molecules can be studied in organs, tissues, or single cells. The properties that make PEG particularly attractive as an embedding medium for *in situ* hybridization studies are the following: (1) PEG is miscible with water, so extensive tissue dehydration, rehydration, and exposure to xylene or toluene is not required; (2) the low melting temperature of PEG requires impregnation temperatures of only 46°C compared to 56 to 60°C for paraffin embedding; and (3) the quality of tissue preservation is generally superior to that obtained with frozen or paraffin sections.

However, PEG has some disadvantages that should be taken into consideration when deciding on an embedding technique for a particular tissue. Perhaps the most challenging problem with PEG-embedded tissue is the mounting of sections onto glass slides (see following section). There might be tissues for which paraffin or frozen embedding is preferable. The method that we describe here for *in situ* hybridization using PEG sections is optimized for brain tissue, but may be adapted to other solid organs.

The histological quality of the tissue processed for *in situ* hybridization is also affected by the procedures used for hybridization. Common protocols for hybridization that require harsh treatments (e.g., exposure to acids, proteases, and ribonucleases) usually result in poor histological quality. In this case, the tissue integrity obtained with the PEG embedding is lost because of the processing that follows. Improved procedures that use none of these harsh steps have been developed.[1] In this chapter we present detailed procedures for *in situ* hybridization that combine the added benefits of PEG with [^{35}S]-labeled riboprobes and minimal tissue treatments.[2] Rather than providing theoretical explanations to the steps involved, we present here a detailed recipe for the procedure.

II. TISSUE FIXATION

Paraformaldehyde at a concentration of 3 to 4% in 0.1 *M* phosphate buffer (pH 7.4) is used. The preferable form of fixation is by intravascular perfusion, although in some cases, immersion fixation may be used (e.g., with small tissue fragments). Here we present a procedure to fix small rodent or bird brains. This technique probably works for other tissues as well.

A. PREPARATION OF THE FIXATIVE

Warm 40 ml of distilled water to 60°C. Add 3 g of powdered paraformaldehyde and stir for 2 to 3 min, maintaining the temperature at 60°C. Add sufficient 5 *N* NaOH (dropwise) until the solution clears. Let the solution cool to room temperature and add 50 ml of 0.2 *M* phosphate buffer (pH 7.4). Adjust the pH of the fixative solution to 7.3 to 7.4 with HCl if necessary, and add distilled water to a final volume of 100 ml. Filter the fixative solution through a 0.45 μm pore membrane.

B. PERFUSION

Anesthesize the animal (we use nembutal, 10 to 30 μg/g body weight) and open the thoraxic cavity to expose the heart. Cut a small hole in the left atrium and inject 0.9% sodium chloride in glass distilled water into the left ventricle until the effluent liquid begins to clear (3 to 5 ml for a canary or 5 to 10 ml for a mouse). Immediately, and with a different syringe, the animal is then injected with the fixative solution (~50 ml total volume).

The tissue to be used is then removed and immersed into the same fixative solution for 0.5 to 2 h at room temperature.

III. IMPREGNATION AND EMBEDDING

The tissue is washed in several changes of phosphate-buffered saline (PBS), pH 7.3, for a total of 1 to 2 h to remove the fixative. PEG is soluble in water, and impregnation methods exist to take the tissue directly from water to PEG.[3] However, we prefer to use an intermediate dehydration step in ethanol before bathing the tissue in PEG. The tissue is immersed in a sequence of solutions of ethanol in water: 50% (1 h), 70% (1 h to 24 h), 95% (1 h), and 100% (45 min). The tissue is then transferred to a series of molten baths of PEG. A stock of PEG 1000 and PEG 1540 (Polysciences, Inc., Warrington, PA) is kept in an oven at 46°C. It is important to keep these PEG stocks well capped, as they will hydrate with time, especially in wet weather. If stock molten PEG becomes hydrated it will result in soft tissue that compresses when is sectioned.

We use a 24-well plate (for tissue culture) to bathe the tissue in a vacuum oven (or a small vacuum desiccator inside an oven) at 46°C in the following baths of PEG: PEG 1000, (1 h); PEG 1000:1540 in a 1:1 mixture, (1 h); PEG 1000:1540 in a 1:1 mixture, (1 h to overnight); and PEG 1540, (10 min).

The times described for dehydration and embedding are for a block of brain tissue approximately 1 × 1 × 1 cm. These times should be increased when the block of tissue is larger or decreased if smaller.

The tissue is placed in the correct orientation in an embedding mold (Peel-a-Way Scientific, South El Monte, CA) filled with molten PEG 1540. The block is then cooled rapidly in a −20°C freezer for 3 min. The hardened block is stored until sectioning at 4°C in an airtight box with desiccant to prevent it from hydration.

The hardness of the embedded tissue may be varied by adjusting the proportion of the different molecular weight PEGs in the mixture: the more PEG 1540 in the mixture the harder the tissue. If the tissue is to be sectioned at less than 5 μm, it is better to infiltrate in a 1:2 mixture of PEG 1000:1540.

However, the best mixture has to be determined empirically for individual applications. The type of tissue, air humidity, and temperature will affect the hardness required for best sectioning.

IV. SECTIONING AND MOUNTING

The PEG block with the embedded tissue is attached to a Lucite® chuck (Carolina Biological Supply Co., Burlington, N.C.) using molten PEG 1540 at 56°C. The tissue is cut in a rotary microtome with a steel knife following standard sectioning procedures as those used for paraffin.[4] Ribbons of PEG sections are easy to generate and are collected with the aid of a hair brush and laid onto rectangular black art board (17 × 39 cm). A second board of the same size is used to cover the sections. In order to prevent the top board from compressing the sections, a 4-mm thick spacer made of rubber foam weather stripping is glued to the perimeter of the lower board. The sections inside the boards are stored with some desiccant at 4°C inside sealed plastic bags.

Mounting of sections onto glass slides requires some practice. This step is perhaps the most challenging one since the PEG is water soluble and the sections disperse from the ribbon when they touch the aqueous mounting solution. Here we describe a method that simplifies the mounting of PEG tissue sections when the cutting surface of the tissue is large (>2 × 2 mm). When the tissue is small or composed of many small fragments, it is generally very difficult to mount the sections onto slides. The tissue fragments disperse on the water surface and separate from each other, as the PEG that holds them together rapidly dissolves. For small fragments of tissue other mounting techniques may be of use[5] (also see other chapters of this book). With paraffin or frozen sections this is not a problem, because the tissue embedment is retained during mounting. These embedding techniques may be more appropriate for tissues that prove too difficult to mount when embedded in PEG.

Sections are floated in a solution of 0.02% diethylpyrocarbonate (to inactivate ribonucleases) in water and are carefully pushed onto the glass slide with a fine paint brush. We have designed a ''section mounter'' (Figure 1) to facilitate this step. It consists of an open plastic reservoir 2.3 × 10 × 5 cm with a black acrylic bottom 1.5 cm from the top. A 45° angled slot on the inner walls of the reservoir and in the black acrylic serves as a ramp for the slide. Two small (1.5 × 0.2 cm) stainless steel channel guides are glued to each side of the ramp. These guides are cut and bent to hold the slide in place by slight spring action. The reservoir is filled with mounting solution. Care must be taken to avoid contamination with ribonucleases (gloves are worn). The sections are cut away from the ribbons and picked up with a 2-pronged fork made from 0.5 mm stainless steel wire. The individual sections are carefully placed to float in the mounting solution and are pushed towards the slide with the brush. The slide is pulled upward out of the mounter, dragging the section along. The section is then positioned and spread with

FIGURE 1. Photograph of "section mounter", fork and small brush. Section mounter is constructed of Lucite®. It holds standard microscope slide at 45° angle next to small rectangular reservoir of aqueous mounting solution. Fork is used to transfer PEG section from the ribbon on to mounting solution. Floating section is driven onto slide with aid of brush. For details see text.

the brush before the mounting medium completely drains from the section. Four to five sections per slide are air dried and stored at 4°C for 1 to 2 weeks. For longer storage the slides are boxed with desiccant at −70°C.

Silanated slides are required to covalently bind the tissue to the slide and prevent the sections from detaching during subsequent incubation. These slides are now commercially available (Digene Diagnostic Inc., Silver Spring, MD), or can be prepared in the laboratory by following a simplified procedure: clean slides in 1 *M* HCl (20 min) followed by 100% ethanol (20 min), then air dry; dip for 5 to 10 sec in a 2% solution in acetone of 3-aminopropyl triethoxysilane (TESPA), (Aldrich Chemical Co., Milwaukee, WI): rinse briefly in acetone (twice), followed by distilled water, and air dry. Coated slides are stable for at least several months. A more elaborate method to silanate the slides that results in better adherence of the section to the slide has been described before.[1]

V. PREPARATION OF PROBES

Hybridization probes compatible with PEG sections include end-labeled synthetic oligonucleotides, double stranded DNA probes, or single stranded RNA probes (riboprobes). Relative advantages of the different types of probes have been discussed elsewhere.[6] Our focus has been on riboprobes, as these probes generally yield higher hybridization signals.[7] However, riboprobes also require harsher, more stringent conditions to reduce nonspecific probe binding to acceptable levels. For studies of moderate-to-abundant RNAs, milder procedures using DNA probes (especially short synthetic oligonucleotides) may result in further improvements in tissue morphology compared to the procedure given below. Our procedure employs ^{35}S as the radiolabel, as this gives an optimal balance between signal strength and spatial resolution. An attractive alternative would be to use recently developed techniques for detection of digoxigenin-labeled probes via conventional high-resolution enzyme histochemical methods (synthesis and detection kits are available from Boehringer-Mannheim Corp. Biochemical Products, Indianapolis, IN), which should complement the improved morphology of PEG sections. However, nonisotopic methods may also not be as sensitive as the one given below. The procedures for probe synthesis and hybridization are similar to the ones described by us previously,[1,2] with several small simplifications or improvements:

Method — Template DNA for riboprobe synthesis (i.e., a cDNA subcloned into a plasmid vector containing SP6, T7, or T3 RNA polymerase promoters, such as pBluescript® from Stratagene, Madison, WI, or pGEM® from Promega, La Jolla, CA) is linearized by digestion with an appropriate restriction nuclease, purified via NENSORB® chromatography as described by the manufacturer (DuPont® Company), and resuspended in water at a concentration of 0.2 mg/ml. A 5× concentrated stock of riboprobe reaction buffer contains the following reagents: 2.5 m*M* each ATP, CTP, and GTP; 0.06 m*M* UTP; 200 m*M* Tris buffer, pH 7.5; 30 m*M* $MgCl_2$; 50 m*M* NaCl, and 10 m*M* spermidine. The riboprobe synthesis reaction is initiated by adding, in order and at room temperature, the following reagents: 3.5 μl alpha-^{35}S UTP (Amersham, Arlington Heights, IL) 400 Ci/mmol in water, 2 μl 5× stock reaction buffer (above), 1 μl 1*M* dithiothreitol (100 m*M* final concentration), 1 μl 1 mg/ml bovine serum albumin (0.1 mg/ml final concentration), 0.5 μl RNAse inhibitor (United States Biochemical or Promega, La Jolla, CA) (minimum 10 units), 1 μl of the purified linearized DNA (0.2 mg, adding more DNA results in shorter probes that often hybridize poorly), and 0.5 μl of appropriate RNA polymerase (T3 or T7 from USB, or SP6 from BRL, Gaithersburg, MD; minimum 7.5 units). The reaction is incubated for 30 min at 37°C, a second 0.5 μl aliquot of polymerase is added, and the reaction is continued for 30 more min. To purify the synthesized probe, the reaction is diluted to 100 μl total volume by addition of diethylpyrocarbonate-treated water[8] and transfer RNA (tRNA) carrier (20 μg); 1 μl is reserved to measure the total radioactivity in the reaction, and the remainder is purified through a Sephadex® G-50 spun column.[8] Typically, 50 to 95% of the radioactivity is incorporated into RNA recovered in the G-50 eluate. The specific activity of the probe is 2.9×10^8 dpm/μg synthesized RNA.

VI. HYBRIDIZATION

Sections (mounted on silanated glass slides) are first acetylated by immersion for 10 min in fresh 100 m*M* triethanolamine and 25 m*M* acetic acid,[9] followed by 3 brief rinses in 2× SSPE (20× stock is 3.6 *M* NaCl, 200 m*M* NaH_2PO_4, pH 7.4, 20 m*M* EDTA), then dehydration in graded alcohols (70, 95, and 100% ethanol) and air drying. Next, they are prehybridized by covering with the hybridization solution minus riboprobe: 2× SSPE, 50% formamide (Fisher Biotech SuperPure® grade), 2 mg/ml each of yeast tRNA and BSA, 1 mg/ml polyadenylic acid, and 100 m*M* DTT. The sections are coverslipped and placed in a 65°C oven for 1 h, after which the coverslips are removed. Excess prehybridization solution may be removed by aspiration, or the sections may be briefly dehydrated. The hybridization solution is made by adding 500,000 dpm riboprobe per 16 μl of solution described above; this is sufficient to cover a section under a 22 × 22 mm coverslip. (Hybridization solution is most conveniently made and stored at a 1.15× concentration, using lab stocks of the components at conventional concentrations and omitting addition of water; subsequent addition of probe and water as necessary will bring the final concentration to 1.0×.) The hybridization solution is applied with a pipetting device. The sections are coverslipped, and the slide is immersed for in paraffin oil at 65°C (e.g., a bath in an oven) for 3 h. The slides are then rinsed with chloroform (to remove oil), then in 2× SSPE (to remove coverslips), and washed as follows: 10 min at 65°C in 50% formamide, 2× SSPE, and 0.1% beta-mercaptoethanol; then twice (20 min each) at 65°C in a low-sodium solution (2 m*M* tetrasodium pyrophosphate, 1 m*M* sodium phosphate buffer pH 7.2, 1 m*M* disodium EDTA, and 0.1% beta-mercaptoethanol; for some probes a less stringent solution of 0.1× SSPE and 0.1% beta-mercaptoethanol may be sufficient).

VII. AUTORADIOGRAPHY

An image of the hybridized section can be obtained most easily by placing it directly against X-ray film (e.g., Kodak® XAR-5, Rochester, NY or Amersham HyperFilm®, Arlington Heights, IL). All but the rarest RNAs (e.g., less than 0.001% fractional abundance in mRNA) can typically be detected after exposures of a few days to a few weeks. For microscopic imaging of labeled cells, slides are dipped in Kodak® NTB-2 emulsion (we typically dilute the emulsion 1:1 with 600 m*M* ammonium acetate prior to dipping, to give a more even coating), followed by standard protocols for emulsion autoradiography.[10] A rule of thumb for exposure time is that it takes about three times longer to obtain an emulsion autoradiograph subjectively comparable in intensity to an XAR-5 film autoradiographic image. After the development of the exposed emulsion, sections can be stained, e.g., using cresyl violet for brain tissue.

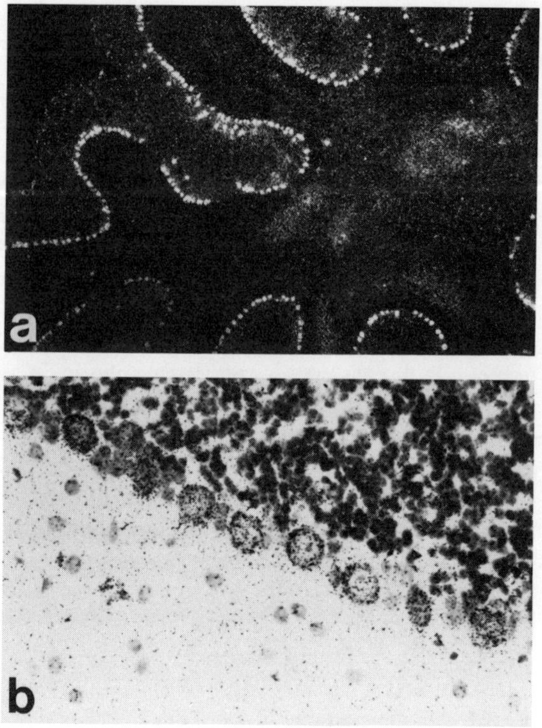

FIGURE 2. Photomicrographs of canary cerebellum section after *in situ* hybridization. Section was prepared and hybridized as described in text, using riboprobe synthesized from cDNA clone "HAT-14" which encodes novel protein kinase C substrate related to GAP-43 and neurogranin (manuscript in preparation). (a) Dark field view of center of cerebellum at low magnification (Magnification × 30). Rows of positive cells (white spots) corresponding to Purkinje neurons, and (b) At higher magnification (Magnification × 300.) and under bright field illumination, autoradiographic grains overlay cytoplasm of Purkinje neurons, but are mostly excluded from large clear cell nuclei.

VIII. CONCLUSION

The morphological detail in PEG sections following *in situ* hybridization procedures is noticeably better than that in frozen or paraffin sections, and signal levels are comparable.[2] An example of a hybridized section following emulsion autoradiography is shown in Figure 2. Cellular detail after *in situ* hybridization in PEG sections allows identification of cell types according to their pattern of Nissl staining. This is very difficult, if not impossible, with frozen sections. Paraffin sections show better morphological detail when compared with frozen sections, but they are not as good as sections prepared using the PEG method. Processes required for paraffin embedding are harsher and more time consuming than for PEG embedding. The technique described here is simple and reproducible. We have been using this method not only for *in situ* hybridization, but for all our routine histology of brain tissue. PEG embedding method should be easy to adapt to many different types of application. However, for some tissues, such as those composed of many small fragments, the production of frozen or paraffin sections may be somewhat simpler, at least initially. Thus each investigator must weigh for each tissue type the relative merits of histological quality vs. simplicity of procedure. PEG sections should be considered as a method of choice to resolve fine cellular details under high microscopic magnification.

REFERENCES

1. **Clayton, D. F., Huecas, M. E., Sinclair-Thompson, E. Y., Nastiuk, K. L., and Nottebohm, F.,** Probes for rare mRNAs reveal distributed cell subsets in canary brain, *Neuron,* 1, 249, 1988.
2. **Clayton, D. F. and Alvarez-Buylla, A.,** *In situ* hybridization using PEG-embedded tissue and riboprobes: increased cellular detail coupled with high sensitivity, *J. Histochem. Cytochem.,* 37, 389, 1989.
3. **Smithson, K., MacVicar, B. and Hatton, G.,** Polyethylene glycol embedding: a technique compatible with immunocytochemistry, enzyme histochemistry, histofluorescence, and intracellular staining, *J. Neurosci. Methods,* 7, 27, 1983.
4. **Humason, G. L.,** *Animal Tissue Techniques,* W. H. Freeman and Co., San Francisco, 1972.
5. **Gao, K. and Godkin, J.,** A new method for transfer of polyethylene glycol-embedded tissue sections to silanated slides for immunocytochemistry, *J. Histochem. Cytochem.,* 39, 537, 1991.
6. **Valentino, K., Eberwine, J., and Barchas, J.,** In situ *Hybridization: Applications to Neurobiology,* Oxford University Press, New York, 1987.
7. **Melton, D., Krieg, P., Rebagliati, M., Maniatis, T., Zinn, K., and Green, M.,** Efficient in vitro synthesis of biologically active RNA and RNA hybridization probes from plasmids containing a bacteriophage SP6 promoter, *Nucleic Acids,* 12, 7035, 1984.
8. **Sambrook, J., Fritsch, E. F., and Maniatis, T.,** *Molecular Cloning,* Cold Spring Harbor Laboratory Press, Cold Spring Harbor, NY, 1989.
9. **Hayashi, S., Gillam, I., Delaney, A., and Terner, G.,** Acetylation of chromosome squashes of Drosophila melanogaster decreases the background in autoradiographs from hybridization with ^{125}I-labeled RNA, *J. Histochem. Cytochem.,* 26, 677, 1978.
10. **Rogers, A. W.,** *Techniques of Autoradiography,* Elsevier/North-Holland, Amsterdam, 1979.

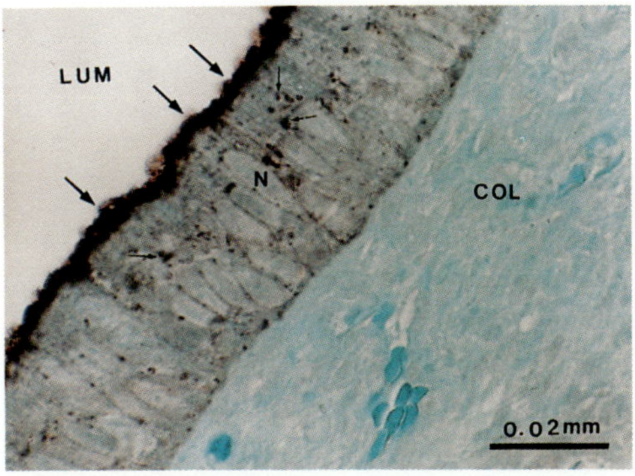

PLATE 1. PAP immunostain of PEG section for localization of TFG-α antigenic site in epithelial cells of bovine uterus. COL, collagen and stromal cells; LUM, lumen of uterus; N, nucleus of uterine epithelial cell; big arrows indicate heavily immunostained TGF-α alone surface of epithelium; small arrows point to vesicle-like structures in cytosol that contain TGF-α before secretion. (Magnification × 625.)

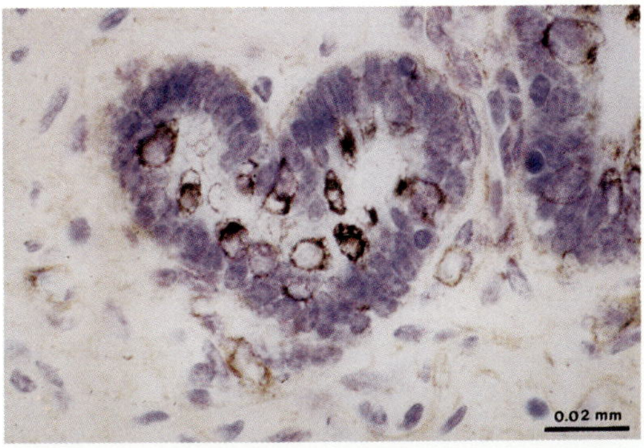

PLATE 2. IGSS of keratin in uterine glandular cells in day-20 pregnant cow. Only those apical parts of glandular cells (very close to lumen of gland) have been heavily immunostained in dark brown. Note the central part of heart-shaped gland. Nuclei are lightly counterstained with methylene blue. (Magnification × 460.)

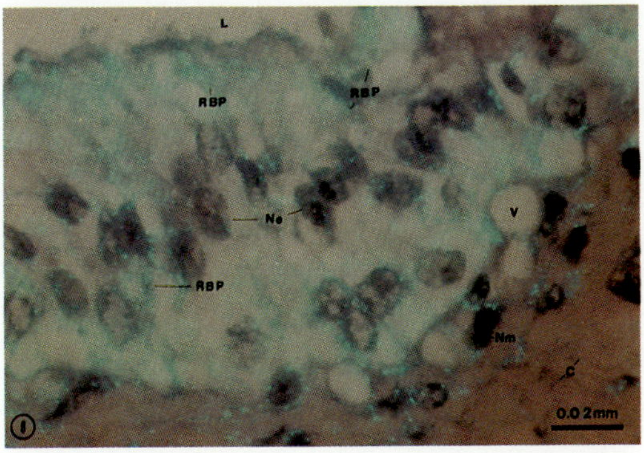

PLATE 3. Localization of bovine placental retinol binding protein in bovine chorionic membrane. IGSS with epipolarization detection plus low intensity transmission illumination. RBP, antigenic sites of retinol binding protein; Ne, nuclei of epithelial cells; Nm nuclei of mesodermal cells; C, collagen fibers; V, vacuol; L, lumen. (Magnification × 390.) (From Gao, K. X., et al., and Godkin, J. D., *Int. J. Develop. Biol.*, 35, 485, 1991. With permission.)

Chapter 7

A NEW METHOD FOR TRANSFER OF POLYETHYLENE GLYCOL SECTIONS ONTO SILANATED SLIDES FOR IMMUNOCYTOCHEMISTRY

Kuixiong Gao

TABLE OF CONTENTS

0-8493-4323-2/93/$0.00 + $.50

I. INTRODUCTION

Paraffin embedding and section preparation methods have been developed and improved throughout the last century. A concise historical review of paraffin and its development as an embedding medium was presented by Sanderson et al. (1988).[30] The paraffin embedding method is still extensively used in biomedical researches and clinical investigations. It may not be overestimated that most of human knowledge for the understanding of animal or plant tissue structures at light microscopic levels is derived from the observations of paraffin sections of fixed and stained tissues. There might not be any other single chemical substance like paraffin, which has contributed so much to the development of histology. Until now, most textbooks related to histology, embryology, physiology, or pathology, etc., still used microphotographs taken from paraffin sections as their basic illustrations.

Because paraffin is water insoluble, any biological materials subjected to the preparation of paraffin sections must go through dehydration and clearing processes, then infiltrate in molten paraffin, which is normally carried out in a thermostatic electric oven with the temperature set just a bit higher than the melting point of the paraffin used. Unfortunately, these steps inevitably require harsh organic solvents, which may cause damage to the enzyme activity or suppress the antigenic reactivity in certain tissues, especially when they are incubated at high temperatures for a long time. These factors constitute an inadequate circumstance for histochemists or cytochemists to use paraffin embedding methods for the localization of delicate enzymes or heat labile antigens. Thus, cryosection has become popular nowadays for histochemistry and immunohistochemistry (IHC) or immunocytochemistry (ICC) (the meanings of these two words are quite close, and they can be alternatively used). Usually, the tissue morphological preservation offered by cryosections is inferior to that of paraffin sections.

Owing to the ever-increasing needs for the study of the macromolecular functions in cells, people are seeking better embedding media that can provide fine morphological features of the cell as well as achieving *in situ* localization of the functional protein (enzyme or antigen) or nucleic acid (DNA or RNA).

Therefore, the polyethylene glycol (PEG) embedding method should be taken into consideration again.

PEGs vary in consistency from liquid to solid, depending on the molecular weight, indicated by a number following the name. Some of the previous references referred to PEG as Carbowax®, which is wax-like in appearance, yet soluble in water, thus differing from common wax or paraffin. PEGs have been utilized in industry and different areas of scientific research. Chapter 1 of this book is an overview of the general physical and chemical properties of PEGs and their various applications.

PEG used as an embedding medium is not a recent story. As early as some 50 years ago, Richards et al. (1942) first successfully produced two electron microphotographs of striated muscle of American cockroach from Carbowax® 4000 section.[27] Blank (1949)[2] used Carbowax® for light microscopic study. These pioneering investigations were followed by other scientists during the last two decades. Throughout 1950s until the 1990s, Blank and McCarthy (1950),[3] Rinehart et al. (1951),[28] Jones and Goodale (1962),[20] Riopel (1962),[29] Johnston et al. (1979),[21] Smithson et al. (1983),[32] Bard and Ross (1986),[1] Cohen et al. (1987),[7] Meininger and Binet (1988),[25] and Gao and Godkin (1991)[11] used PEG as an embedding medium for light microscopy (LM). During this period, PEG and its derivatives were also used as a removable embedding material for electron microscopy (EM).[4,5,22,23,33] The PEG embedding method was developed rather slowly and has not yet been accepted by the majority of histochemists and cytochemists. Perhaps the main reason that people do not use PEG as an embedding material for microscopy is the difficulty in handling PEG sections, which tend to sink into water or buffered solutions. The transfer of PEG sections may be tricky. Supposedly, sometimes people feel it is difficult to adequately repeat the procedures for transfer of PEG sections as reported by the original papers.

In any case, there are several transferring techniques for PEG sections with detailed descriptions.[2,24,26,29] Based on these techniques, satisfactory results from PEG sections of different types of tissues were obtained, including plant materials,[18,29] monoclonal antibody screening,[1] and molecular probe *in situ* hybridization.[6] Furthermore, Gao and Godkin (1990)[11] have recently introduced a new method using the agarose blocks for transferring PEG sections onto silanated glass slides for immunocytochemistry, which overcomes the most annoying problem in dealing with PEG sections. In this chapter, I would like to describe that new method in more detail and to share some of my personal experiences in immunostaining PEG sections.

II. SELECTION OF MOLECULAR WEIGHT OF PEG FOR IMMUNOCYTOCHEMISTRY

A. PEG 400

Theoretically, the lower the temperature in processing tissue samples, the less harmful to the antigenic reactivity it will be. In order to cut sections, tissue must be supported by surrounding materials that aid in cutting. PEG 400 is a liquid at room temperature. Therefore, it is not suitable for use as an embedding material for cutting sections at room temperature. However, PEG 400 turns into solid when it cools below its melting point. Mowery et al. (1989) developed a new method for PEG 400 embedding and sectioning.[26] They claimed that PEG 400 was an ideal embedding medium for immunocytochemistry when the blocks are cut with a cryomicrotome. (See also Chapter 3 of this book.)

B. PEG 1540

PEG 1540 is a waxy solid at room temperature. The hardness of the solidified block derived from PEG 1540 is moderate, which is very good for cutting sections of tissues embedded within it. The melting point of PEG 1540 is around 55°C, a temperature that will not cause great trouble for moderately rigid antigens in retaining their immunoreactivities after embedding. There will be no trouble at all in obtaining section ribbons sliced from a PEG 1540 block with a common rotary microtome, if the operator has experience in making paraffin sections. The section thickness may be adjusted, ranging from 3 to 10 μm. For the purpose of ICC or enzyme localization, PEG 1540 is really the first choice among all PEGs, because employing PEGs with higher molecular weight requires incubation of tissues at a higher temperature equivalent to the melting point of the specific PEG used. As a general rule, high temperature is unfavorable to most enzymes and antigens.

In this chapter, all the methods are applied for embedding tissue samples with PEG 1540 alone; however, the principle may extend to other PEG sections. The high molecular weight of PEG for EM will not be a concern here. Readers can learn these techniques elsewhere in this book.

III. EMBEDDING AND SECTIONING WITH PEG 1540

A. PURIFICATION OF PEG 1540

The commercially available PEG 1540 is a waxy solid mass. Polysciences Inc. (Warrington, PA) provides its PEG 1540 in the form of a large piece of wax-like block whole in a plastic bottle. It takes several hours or overnight incubation in an oven at 55°C or higher to melt the PEG 1540 block before use. For convenience sake, it is suggested to divide the molten PEG into small quantities for storage in congealed form. The molten PEG 1540 can be mixed with any portions of warm water to form different concentrations. When the concentration of PEG is higher than 50%, it will form a semi-solid mass at room temperature; below that concentration the PEG 1540 solution remains a thick liquid at room temperature. The newly purchased PEG 1540 can always be used directly without further purification, although it may have slightly acidic smell. PEG 1540 is hygroscopic. When the bottle is not tightly sealed, the moisture will get into the PEG quickly, or, if the PEG was stored in a higher temperature, such as in an oven, the lower molecular weight PEGs will be produced due to the self-cleavage. Such types of wetted and decomposed PEG will definitely cause trouble for the final embedding, i.e., unevenness of the texture of the block with cracks, whitish weak points, or multi-bead-like structures, which are very bad for sectioning. When this is the case, it seems that purification is unavoidable before starting embedding process. An economic way of purifying wetted and decomposed PEG 1540 is quite simple and is very important in achieving high quality research work. The purification procedures are as follows:

1. Put the wetted PEG that needs to be purified in an oven at 60°C until all the solid parts are melted thoroughly.
2. Move the container (or the bottle with a lid) to another oven, setting the temperature at 37 to 40°C.
3. Wait a few hours to let the temperature drop; the PEG 1540 will congeal and accumulate around the wall of the container, retaining a certain thick liquid form of low molecular weight PEGs with moisture in it. This liquid form of PEGs should be poured out and discarded.
4. Put the container with the solid PEG back in a 55 to 60°C oven to help remelt the partially purified PEG 1540 for use in experiments.

Normally, one purification is enough to bring back the quality of PEG 1540 for satisfactory embedding and sectioning. However, a second run of steps one through four will be even more helpful for obtaining PEG sections of superior quality.

PEG 1540 is actually a mixture of PEGs with an average molecular weight of 1540 Da. Because the manufacturers may not always keep the exact quality of PEG products in every lot, I would rather purify the newly purchased PEG each time. The purified PEG 1540 is always better to use. I do not add high molecular weight PEG, such as PEG 4000 to PEG 1540 to adjust the hardness of the block. The cutting property of the purified PEG 1540 is perfect in most cases.

B. INFILTRATION AND EMBEDDING

1. Infiltration of Solid Tissue Slice

PEG 1540 can mix with water or alcohol freely. The tissue slices, after fixation in 1 to 4% paraformaldehyde in 0.1 *M* phosphate buffer, pH 7.4, (or Bouins fixative, or any other preferred fixative solution routinely used by the investigators), are washed extensively with distilled water or buffered saline, to remove residuals of fixatives. Then they are treated with a permeabilization solution. The permeabilization solution consists of the following reagents: Surfact-Amps® X-100 (Pierce, Rockford, IL), (1 ml); NH_4Cl (1 g); 2% NaN_3 (1 ml); and 0.45% NaCl in 0.025 *M* phosphate buffer, pH 7.4 (100 ml). Time needed for treatment depends on the size of the tissue being treated. For instance, a tissue slice thinner than 2 mm is treated for 10 to 15 min.

After permeabilization the tissue slices are infiltrated with 10, 20, 30 and 50% of PEG 1540 in distilled water at room temperature, each step for 20 min, or till the samples sink to the bottom of the vessel that contains the PEG solution. Then the tissues are further incubated in 70, 80, and 90% PEG 1540 in water, followed by two changes of 100% PEG 1540 in an 55°C oven, each step for 20 min or longer. Samples are then embedded with molten PEG 1540 and cured in the aluminum (or other metallic) embedding mold.

To obtain a nice block for easy cutting, touch the bottom part of the mold to the ice bath, which can decrease the temperature of molten PEG effectively (compared with paraffin, the molten PEG is slower to cool down). A rapid chilling results in more consolidation of congealed PEG 1540 facing the bottom of the mold where the tissue is placed. Then, the block is allowed to cool in a refrigerator. It is not good to put the embedding mold in a freezer at −20°C. This will cause cracks to form.

To avoid difficulty in peeling off the solidified PEG block from the metallic mold, it is wise to smear a thin layer of Vaseline® or mineral oil on the surface of the mold beforehand. By doing this, a nice PEG block can be safely separated from the mold without damage to the tissue. Blocks of high quality are beautifully homogenous, semitranslucent, and without whitish small flaws.

2. Infiltration and Embedding of Delicate Embryonic Membrane

Embryonic tissues contain more water than the adult ones. Amnion, allantois, and chorionic membrane of early developmental stages are extremely delicate. These membranes are thinner than a sheet of lens tissue. If they are fixed in a solution without any support, they will twist together and cause misunderstanding what would correspond to the real outer (or inner) surface in living state. Herewith, I introduce a simple solution. The specific embryonic membrane is collected in a small plastic petri dish containing phosphate-buffered saline (PBS) and a base of silacone gel prepared with 184 Silicone elastomer kit (Dow Corning, Midland, MI), Silgard® which is transparent and highly elastic. Use micro insect pints (stainless steel, 10 mm long, 0.15 mm in diameter) to fix the membrane over the Silgard® gel. It is best to work under a dissecting binocular microscope. Then, remove the PBS and fix the membrane in the same dish with suitable fixatives, such as 4% paraformaldehyde in PBS, pH 7.4, for 1 h at room temperature. Then follow by washing, permeabilization, and infiltration steps, all in the same dish without disturbing the membrane, until it goes to the 100% PEG 1540 (similar to the procedures in Section III, B.1. At that time the membrane hardens and can be cut to proper size in the melted PEG, which can be transferred to an aluminum mold for the final embedding.

3. Infiltration and Embedding of Liver Tissue

Liver, which contains a large amount of polysaccharide and lipid, is a kind of tissue known to be rather difficult to cut evenly when it is embedded with paraffin. Evidently, the normal procedure for the infiltration and embedding of common tissues is not quite suitable for fixed rat liver. I found that totally dehydrated rat liver in a PEG 1540 block was very difficult to cut, because the sections will break into debris, while a partially hydrated liver can be cut much more easily. If the liver tissue is infiltrated in a series of graded increased concentrations of PEG in water, stop it in a final incubation medium composed of 5% glycerol and 5% water in PEG 1540, then directly embed it in PEG 1540. With the liver tissue prepared in such a way and cut only in a highly moisturized surrounding (see next section) it is possible to obtain serial section-ribbons.

C. SECTIONING AND SECTION-RIBBON STORAGE

The majority of the authors using the PEG embedding method claimed that cutting PEG sections is best done dry, preferably in an air-conditioned room. Ribbons of PEG section (5 μm or thinner or thicker) are really as easy as cutting paraffin sections with a rotary microtome. A highly humid surrounding is not suitable for cutting PEG sections. The main difficulty is that the moisturized section-ribbons tend to stick to the cutting edge of the knife or to the filter paper that collects them. However, I would like to add few words here to remind you that some exceptional cases exist.

If the weather is extremely dry, static electricity builds up easily. This will be obvious after continuous cutting, especially before a big thunderstorm when the air may be full of static electric charges. The overdried air facilitates the accumulation of electric charges, which may cause attraction or flying of the ribbons. On the other hand, cutting in a moderately humid surrounding is much better. To clean up the charged fragment, I recommend using the Staticmaster® (NRD, Inc., Grand Island, NY) which is a brush with polonium radiation that reduces static electronic charge.

Another extraordinary circumstance is cutting PEG-embedded liver. I have tried to cut liver in dry air but failed. No matter how carefully I cut, the liver sections still broke into debris. At last I learned that only a very, very highly humid surrounding would favor obtaining section-ribbons of PEG-embedded liver tissue. Sometimes I have to below more moisture onto the surface of the cutting block with a plastic pipette before every cutting to ensure the quality of rat liver sections.

The PEG section-ribbons are collected on a sheet of clean filter paper. They must be stored in a desiccator. It is most suitable to use a tinted plastic desiccator cabinet of 25 × 25 × 25 cm with 4 sets of shelves and a front door (Electron Microscope Sciences, Fort Washington, PA). To avoid static electricity that will cause annoying problems of flying ribbon in the plastic desiccator, do not rub the surface of the plastic container. This will produce unwanted charges. For safety reasons, one may spray the outside of plastic desiccator with commercial electrostatic charge neutralizer — the Chemtronics Static Free® (Chemtronics Inc., Kennesaw, GA).

D. TRANSFER OF PEG SECTIONS WITH MINUTE AGAROSE BLOCKS

Transfer of PEG sections with minute agarose blocks onto Digene® silanated slides (Digene Diagnostics, Inc., Silver Spring, MD) for immunohistochemistry was introduced by Gao and Godkin (1990),[11] and Gao et al. (1991).[12] That method was further improved through my personal practice. To help readers to apply this technique for their own purposes, I will also deal with some processes in making tools not yet commercially available.

1. Preparation of Minute Agarose Blocks

The first step is to weigh 3 g of low-gelling temperature agarose (either Sea Plaque® agarose from FMC Corp., Rockland, ME, or, Amresco® agarose I from Amresco, Solon, OH). Pour the agarose powder into a Pyrex® Erlenmeyer flask containing 100 ml of distilled water. Heat the flask in a boiling water bath to dissolve the agarose with gentle stirring. An electric heating plate with a magnetic stir bar works well. Then add 1 ml of 2% NaN_3 to the agarose sol (which may need to be degassed if too many bubbles form). If evaporation during the time of preparation is not considered, this is roughly a 3% agarose sol. Pour 35.3 ml of the agarose sol into a 15 × 150 mm plastic petri dish, tilt the dish in different directions to help spread the agarose sol all over the bottom surface while the gel is still warm. Then immediately float the agarose-containing dish over the surface of cold water, about 4 to 10°C, in a glass pan. Do not disturb the floating petri dish for 10 to 15 min. This permits the agarose sol to turn into a very flat gel layer about 2 mm thick. With the help of a scalpel and guided by a ruler, it is easy to cut smooth cross lines into the agarose gel. Minute agarose blocks of desired sizes can be made without special training. A block of 2 × 16 × 16 mm in size is the most user-friendly for PEG section transferring. Store the minute agarose blocks in a small plastic box or a capped bottle filled with distilled water (containing 0.02% NaN_3 as an antiseptic agent).

2. How to Transfer PEG Sections With Minute Agarose Blocks

First, place a piece of minute agarose block on a small piece of thick filter paper. Then wipe away excess storing fluid. Next, add 10 to 20 μl of flotation medium (flotant) onto the top of the agarose block with a pipette. The flotant is composed of pH 7.4 PBS with 3% PEG 200 and 0.02% Surfact-Amps® X-100. Wait for a while until the liquid spreads over the top surface of the minute agarose block. Cut the section ribbon into small fragments (two to three sections) with a sharp scalpel, then use a pair of watchmaker's forceps to pick up a piece of the fragment at one of its corners and move it horizontally to the upper position over the prepared agarose block. Lay the PEG sections flat in order to touch the thin layer of flotant on the top of the agarose block. The PEG dissolves from the tissue sections into the flotant instantly. Because the surface of the agarose block is relatively small, the floating sections lie quietly and undistorted over the flotant. Then move the section-loaded agarose block together with its filter paper base to a moisture chamber and let the flotant and dissolved PEG be absorbed into agarose.

When the expensed sections fully attach to the surface of the agarose block, they look slightly opaque over the shiny agarose surface. They are in a transitional form of drying, but are still wet, so there will be no drying effect to damage cellular structures. Now, lie a piece of Digene® silanated glass slide on a bench top, and add a drop of 5 μl of distilled water to its surface. With the aid of watchmaker's forceps again, flip the section-loaded minute agarose block upside-down on the Digene® slide in a manner similar to the section-mounting procedure with a coverslip. The PBS drop on the slide will help achieve a blotting of agarose block on the slide without bubbles between them. An intimate contact between the sections and the slide is established when the intermediate liquid is absorbed into the agarose block. After sections are blotted onto it for 10 to 15 min at room temperature, the slide is then inserted into a Coplin jar containing PBS. The agarose block will automatically separate from the slide but leave the sections being printed onto the slide through the covalent bond binding. Now the sections are ready for immunostain.

IV. IMMUNOCYTOCHEMICAL STAIN OF PEG SECTIONS FOR LIGHT MICROSCOPY

A. MAGNETIC OPERATING MICROBRIDGE

Although PEG sections can be immunostained in a way similar to paraffin staining in Coplin jar, it would be much better to use microchamber method to increase the quality of immunostain, and meanwhile, to save prestige of the antibody.[11,12,14] A magnetic operating microbridge is designed for this purpose.

The magnetic operating microbridges are not yet commercially available. They have to be handmade by the investigators. The first model of magnetic microbridge is easy to make according to the description of Gao and Wei (1990).[14] A polycarbonate plastic sheet 1 mm thick is cut into small rectangular pieces of 12 × 14 or 16 × 18 mm, depending on the size of the sections to be tested. Then, with dissolved polycarbonate in dichlorodimethylethane, or simply with nail polish, glue two needle-like iron-steel bars to each of the small rectangular plastic pieces at their two longitudinal ends, but on the same side. The diameter of the metallic bar is about 0.75 mm, while the length of it may vary from 12 or 16 mm, depending on the width of the plastic piece selected. All the surfaces of the iron-steel bars should be painted with the coating material to form a thin layer of the glue or the nail polish to prevent the iron-steel from rusting when in contact with salt solution during immunostaining. An enlarged diagram of a magnetic-operating microbridge is shown in Figure 1. Because magnetic-attractive material is incorporated in

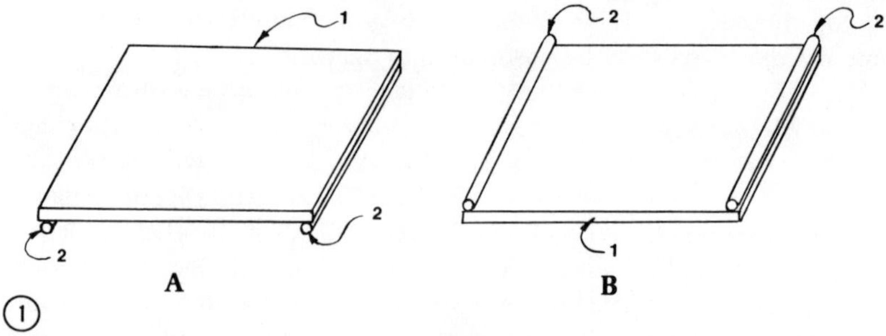

FIGURE 1. Magnetically attractive microbridge, Model-I. (A) Oblique view; (B) Upside-down position. Two iron-steel rods about 0.75 mm in diameter (2) are glued to small piece of polycarbonate plastic sheet (1) at its two ends. (Magnification × 1.8.)

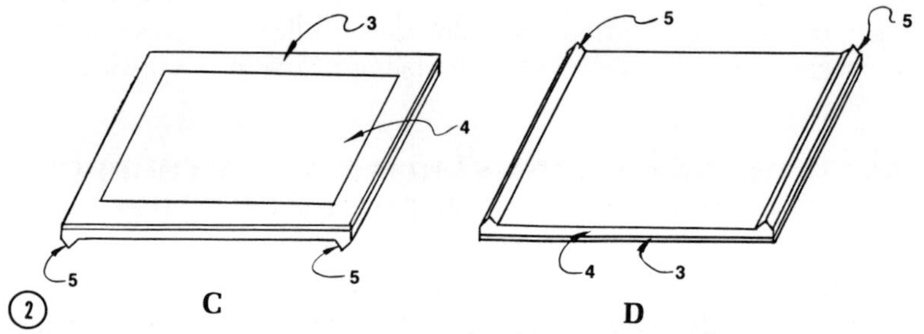

FIGURE 2. Magnetically attractive microbridge, Model-II. (C) Oblique view; (D) Upsidedown position. Minute frame made with iron-steel or other magnetic attractive material (3) is incorporated in fabricated plastic microbridge (4) with two tracts at same side near its two ends (5). (Magnification × 1.8.)

the microbridge, it will be held firmly on the surface of the slide by the magnetic force generated by a magnetic slide placed underneath the section-loaded slide (see the model in Figure 3), as it is placed onto a section-loaded slide. Thus, a temporary microchamber can be built over the slide, with which the medium needed to use in the process of immunostaining can be introduced into the restricted area on the slide and spread evenly over the sections to be stained.

Alternatively, a newly designed model of magnetic operating microbridge incorporates a minute iron frame into the top of a fabricated plastic bridge, as indicated in Figure 2. The benefit of this model is obvious, because it has magnetic-attractive materials all around the edges of the microbridge. One may expect models with iron-steel frames would offer a firmer grounding onto the glass slide by the magnetic slide underneath (Figure 3). Such a model might have marketing value if mass production could be accomplished economically.

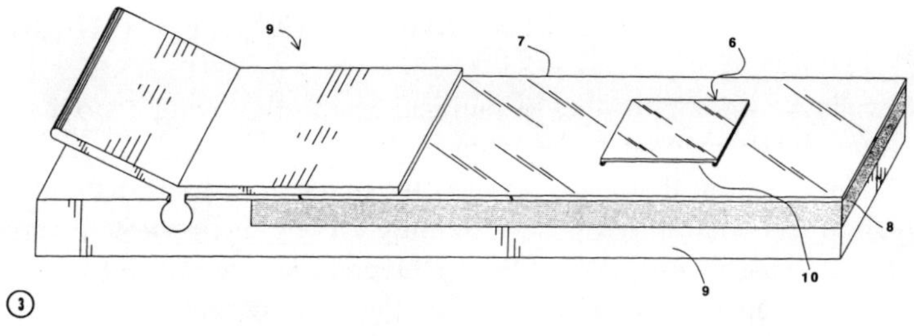

FIGURE 3. Magnetic slide and its holder. Magnetically attractive microbridge (6) is placed on glass slide (7), which is laid over magnetic slide (8) with holder (9), which is composed of two joint parts: base plate and clipper. Microbridge is secured on surface of slide to form microchamber (10) for holding medium for immunostaining. (Top-side view, magnification about × 0.8.)

B. IMMUNOSTAIN IN THE MICROCHAMBER

Place a piece of magnetic-attractive microbridge onto the top of a slide above the sections to be stained. The microbridge is secured in place by the magnetic force generated from a magnetic slide underneath a glass slide. A temporary microchamber is established between the microbridge and the surface of the slide. The incubation medium is introduced into the microchamber through one of its two openings with a pipette. The liquid is pulled into the microchamber by capillary force, and it is sucked away from the microchamber automatically with a pipette connected to an inspirator. The washing steps can be performed in a similar manner, or in a Coplin jar.

The benefits of applying the microchamber immunostain method are as follows:

1. It can save antiserum. Normally, the volume of diluted antibody solution to fill is as little as 80 to 120 μl (according to the size of the microbridge selected).
2. Not only can the microbridge reduce the volume of antibody, it can also provide an even distribution of staining solution all over the staining area. It is unlikely to drop solution on a slide, which has a meniscus and tends to spread.
3. It also can prevent the sections from drying, thus decreasing the background stain.
4. It permits the investigator to examine the sections anytime during immunostaining or coloration.
5. The microbridge is removable. It is feasible for counterstaining and final slide mounting.

Several trials of immunostain with the microchamber method proved it to be successful. The resulting microphotographs are excellent.[11,12]

C. IMMUNOSTAIN WITH PEROXIDASE-ANTI-PEROXIDASE METHOD

Nowadays, immunohistocytochemical stain employing Peroxidase-anti-Peroxidase (PAP) method is so extensively used in biological research and clinical investigations that it has become one of the basic tools in immunostain. The method has been developed as staining kits by different companies. For example, the Histo Gen® PAP Rabbit Universal Kit produced by Bio Genex Laboratories, San Ramon, CA is one of the most reliable PAP-stain kits of its kind that I have personally used for many years. It is beyond the scope of this chapter to refer to the vast practices with PAP method by different immunohistochemists in different areas of investigations. The antigens I have tested include keratin, retinol-binding protein, TGF-α and TGF-β1, and several antigenic enzymes, etc. The primary antiserum raised against TGF-α or TGF-β1 is very weak. Actually it is only against a fragment of the antigen mentioned, per se 13 amino acids. The same antiserum is too weak to detect the antigenic site in tissue with paraffin-embedded sections using the similar procedure of PAP staining. On the contrary, it showed sharp results when PEG 1540 sections were stained with PAP method, implying that PEG sections are superior to paraffin sections regarding the localization of weak antigenic sites. Plate 1* is a micrograph of PAP stain of TGF-α in bovine uterus. However, I feel that immunogold plus silver enhancement will give even higher signals for the detection of the same antigenic site in the same tissue with the same primary antibody at the same concentration. The method will be described in the following section.

D. IMMUNOGOLD AND SILVER ENHANCEMENT TECHNIQUE (IGSS)

Immuno-gold stain followed by silver enhancement, or, immunogold-silver stain (IGSS) is hitherto one of the most sensitive immunological staining methods for detecting antigenic sites in cells or matrix, because the signals are greatly amplified by silvering of the invisible colloidal gold particles (which are in the range of 1 nm to 15 nm, below the particle size discernible in the microscope at its highest magnification) to visible ones.[17] Silver enhancement enables the gold particles to be seen in the normal LM, but even better in the epipolarized microscope, which we will discuss later.

The procedure of IGSS that I have been using is quite similar to the method described by other authors.[8,9,16,17,19] However, I preferred to use a low temperature (with ice bath to control the temperature around 10 to 15°C). This can decrease the nonspecific background. The procedure and results of IGSS on PEG sections using microchamber immunostain method have been published.[11,12] It is briefly quoted in the following:

1. PEG sections are transferred onto the Digene® silanated slide using

* Plate 1 follows numbered page 79.

minute agarose block transferring method and applied to immunostain in microchamber as indicated in the previous section.

2. Sections are permeabilized with 1% Surfact-Amps® X-100 and 1% NH_4Cl in half-strength PBS, pH 7.4, for 15 min.
3. Blocking in a blocking medium (1% bovine serum albumin, 5% normal goat serum, 0.1% cold water fish gelatin from Sigma Chemical Co., St. Louis, MO, in PBS containing 0.02% NaN_3) for 45 min or longer at 37°C, or overnight in a refrigerator.
4. Incubation in medium containing primary antibody (rabbit antiserum) with appropriate dilution at 37°C for no less than 1 h (or 3 h at room temperature, or overnight at 4°C). The incubation medium may use a solution equal to the blocking medium, or reduce its amount of goat serum to 1%. Use the normal rabbit serum of the same dilution to the first antibody as a control.
5. Suck out the primary antibody solution and rinse with cold PBS followed with wash in cold PBS in a Coplin jar no less than 3 to 5 min.
6. Incubation in 1:30 dilution of 1 nm gold-labeled goat anti-rabbit IgG (Janssen Life Sciences Products, Olen, Belgium, or, similar product from other companies) in incubation medium at room temperature for 45 min or longer.
7. Wash in PBS as in step 5.
8. Fix the sections in 2% glytaraldehyde at room temperature for 5 to 10 min followed by wash in glass distilled water.
9. Silver enhancement using a freshly prepared mixture of IntenSE™ M (Janssen Life Sciences Products, Olen, Belgium, or, similar product from other companies) at 10 to 15°C for 6 to 12 min.
10. After counterstain, or without counterstain, the specimen is mounted in an aqueous mounting medium or other mounting medium according to availability.

V. HIGH SENSITIVE EPIPOLARIZED MICROSCOPY DETECTION

A. FOR PEG THICK SECTIONS

Plate 2* is a sample of a typical IGSS. The image of immunoreactive site is dark brown in color. IGSS is a very sensitive method. In most cases, the IGSS method is satisfactory for research, even for moderately difficult projects. In cases when the antigen being tested is extremely delicate and not easy to preserve even by PEG embedding method, or the concentration of the antigen is too low and only marginally detectable, the use of epipolarized microscopy seems rather helpful in acquiring convincing data. As mentioned by several authors, the epipolarization detection of immunogold-silver enhanced particles may increase the sensitivity up to about 20-fold more than

* Plate 2 follows numbered page 79.

a conventional LM observation.[8,9,19] Another gain in employing epipolarization technique is that a low-level intensity of transmitted light can be mixed with epipolarized illumination to give out the shiny brilliant immunostain of antigenic site with a combination of the tissue profile at the same picture. The result is shown in Plate 3*. Readers are referred to the microphotographs in References 9, 11, and 12 as well.

B. THIN SECTIONS PREPARED FROM PEG SECTIONS AFTER IGSS

There is no doubt that IGSS of PEG sections is already good enough to solve problems for the identification of the presence of the questioned antigen within the tissue. But if higher resolution for a more accurate subcellular localization of the antigenic distribution is critical, the 5 μm PEG sections after completing the IGSS may further "en bloc" counterstained with commonly used dye solution, and dehydrated and reembedded in Visio® Bond, which is a blue light sensitive and instantly curable resin.[13,15] It takes 20 to 40 s under the illumination with a light "pistol" to cure the resin. The reembedded PEG section is then mounted onto a blank Epon block with Visio® Bond again, which is cut with a sapphire knife on an ultramicrotome to obtain semi-thin sections of 0.5 μm. The semi-thin section obtained is only one tenth of the thickness as compared to that of the original PEG sections. The Visio® Bond semi-thin sections are collected with a hair loop and applied to a clean microscopic slide. After warm drying on an electric plate at 60 to 80°C for 2 to 3 min, they are coverslipped with the aid of Visio® Bond. Undoubtedly, the resolution of Visio® Bond sections would be increased tremendously, because that they are 10 times thinner than 5 μm PEG sections. With that method, the localization of glycogen metabolic-related enzymes (i.e., phosphoenol pyruvatekinase, glycogen phosphorylase, and glycogen synthase) in hepatocytes were revealed much better in comparison with that produced with 10 μm cryosections (see micrographs in Reference 10). Color pictures taken from semi-thin sections of Visio® Bond-embedded specimens were shown with combined image of both antigenic sites by IGSS (with epipolarized illumination) and the cellular profile (with normal transmitted illumination).

VI. CONCLUDING REMARKS

PEGs have been used as a water-soluble embedding medium for both LM and EM for about 50 years. One limiting factor that prevents it from routine use is the difficulty in transferring PEG sections to slides. The newly developed method using agarose block to transfer PEG-embedded tissue sections onto silanated glass slides has duly solved that problem. With the increasing

* Plate 3 follows numbered page 79.

needs of highly sensitive detection of weak antigenicity in cells, the water soluble PEG embedding method is predicted to be welcomed by most immunohistocytochemists. Hopefully, the methods described in this chapter will be more and more familiar to researchers in different areas of investigation.

REFERENCES

1. **Bard, J. B. L. and Ross, A. S. A.,** Improved method for making high affinity sections of soft tissue embedded in polyethylene glycol (PEG): its use in screening monoclonal antibodies, *J. Histochem. Cytochem.*, 34, 1237, 1986.
2. **Blank, H.,** A rapid embedding technique for histologic sections employing a water soluble wax, *J. Invest. Dermatol.*, 12, 95, 1949.
3. **Blank, H. and McCarthy, P. L.,** A general method for preparing histologic sections with a water soluble wax, *J. Lab. Clin. Med.*, 36, 776, 1950.
4. **Capco, D. G., Krochmalinie, G., and Penman, G.,** A new method of preparing embedment free sections for transmission electron microscopy: applications to the cytoskeletal framework and other three dimensional networks, *J. Cell Biol.*, 98, 878, 1984.
5. **Capco, D. G. and McGranghey, R. W.,** Cytoskeletal reorganization during early mammalian development: analysis using embedment free sections, *Develop. Biol.*, 115, 446, 1986.
6. **Clayton, D. F. and Alvarez-Buylla, A.,** *In situ* hybridization using PEG embedded tissue and riboprobes: increased cellular detail coupled with high sensitivity, *J. Histochem. Cytochem.*, 37, 389, 1989.
7. **Cohen, E., Binet, S., and Meininger, V.,** *In situ* appearance of the cold stable microtubules in the growing axons of the tectal plate of mouse investigated immunocytochemically after polyethylene glycol (PEG) embedding, *Brain Res.*, 433, 171, 1987.
8. **De Waele, M., De Mey, J., Renmans, W., Labeur, C., Jochmans, K., and Van Camp, B.,** Potential of immunogold-silver staining for the study of leukocyte subpopulations as defined by monoclonal antibodies, *J. Histochem. Cytochem.*, 34, 1257, 986.
9. **De Waele, M., Renmans, W., Segers, E., De Valck, V., and Jochmans, K.,** An immunogold-silver staining method for detection for cell surface antigens in cell smear, *J. Histochem. Cytochem.*, 37, 1855, 1989.
10. **Gao, K., Morris, R. E., Giffin, B. F., and Cardell, R. R.,** Immunocytochemical detection of hepatic carbohydrate metabolic enzymes: improved resolution with polyethylene glycol embedding and Visio-Bond semithin sections, *Proc. EMSA*, (50), *MAS*(27), *MSC*(19) *Annu. Meet., August 16-21, 1992, Boston, MA*, Part I, Bailey, G. W., Bentley, J., and Small, J. A., Eds., San Francisco Press, 1992, 792.
11. **Gao, K. and Godkin, J. D.,** A new method for transfer of polyethylene glycol-embedded tissue sections to silanated slides for immunocytochemistry, *J. Histochem. Cytochem.*, 39, 537, 1991.
12. **Gao, K., Liu, K. H., and Godkin, J. D.,** Immunohistochemical localization of bovine placental retinol-binding protein, *Int. J. Dev. Biol.*, 35, 485, 1991.
13. **Gao, K., Smith, S. E., and Godkin, J. D.,** Phagocytosis of protein coated colloidal-gold-agarose-gelatin microbeads by cultured uterine glandular epithelial and stromal cells, *Biotechnol. Histochem.*, 1, 1, 1991.
14. **Gao, K. and Wei, C.,** Gao's microchamber method for immunohistochemistry, *J. Cell Biol.*, 111(5, part 2): 57a(301), 1990.

15. **Gao, K. and Peng, B.,** Visio-Bond as an embedding medium for light and electron microscopy, *J. Cell Biol.*, 105 (Suppl. 4) 226a (1278), 1987.
16. **Hayat, M. A.,** *Colloidal Gold: Principles, Methods, and Applications,* Academic Press, San Diego, 1989.
17. **Holgate, C. S., Jackson, P., Cowen, P. N., and Bird, C. C.,** Immunogold-silver staining: new method for immunostaining with enhanced sensitivity, *J. Histochem. Cytochem.*, 37, 938, 1983.
18. **Holubowicz, R. and Goffinet, M. C.,** Polyethylene glycol embedment for histological studies of bean seed testa of low moisture content, *Stain Technol.*, 63, 33, 1988.
19. **Jackson, P., Lewis, F. A., and Wells, M.,** *In situ* hybridization technique using an immunogold-silver stain system, *Histochem. J.*, 21, 425, 1989.
20. **Jones, R. M. and Goodale, F.,** Improvements in the method of embedding tissues in Carbowax®, *Am. J. Clin. Pathol.*, 38, 173, 1962.
21. **Johnston, M. C., Noden, D. M., Hazelton, R. D., Coulombre, J. L., and Coulombre, A. J.,** Origins of avian ocular and periocular tissue, *Exp. Eye Res.*, 29, 43, 1979.
22. **Kondo, H.,** Correlative light, scanning, and transmission electron microscopy of polyethylene glycol (PEG) embeded and subsequently deembedded tissues, Bailey, G. W., Ed., *Proc. 4th Annu. Meet. Electron Microscopy Soc. Am.*, San Francisco Press, San Francisco, 1989.
23. **Kondo, H.,** Visualization of the inner and outer surfaces of the cell membrane and cytoskeleton by polyethylene glycol embedding, subsequent de-embedding, and rotary replication with platinum, *J. Electron Microsc. Tech.*, 7, 17, 1987.
24. **Mazurkiewicz, J. E. and Nakane, P. K.,** Light and electron microscopic localization of antigens in tissues embedded in polyethylene glycol with a peroxidase-labeled antibody method, *J. Histochem. Cytochem.*, 20, 969, 1972.
25. **Meininger, V. and Binet, S.,** Spatial organization of microtubules in various types of cells in the embryonic tectal plate of mouse immunofluorescence after PEG embedding, *Biol. Cell.*, 64, 301, 1988.
26. **Mowery, J., Chener, J., Spangenberger, S., and Hixson, D. C.,** Rapid low molecular weight polyethylene glycol embedding protocol for immunocytochemistry, *J. Histochem. Cytochem.*, 37, 1549, 1989.
27. **Richards, A. G., Anderson, T. F., and Hance, R. T.,** A microtome sectioning technique for electron microscopy illustrated with striations of striated muscle, *Proc. Soc. Exper. Biol. Med.*, 51, 148, 1942.
28. **Rinehart, J. F. and Abul-Haj, S.,** Histological demonstration of lipids in tissue after dehydration and embedding in a polyethylene glycol, *Arch. Pathol.*, 51, 666, 1951.
29. **Riopel, J. L.,** Carbowax® for embedding and serial sectioning of botanical material, *Stain Technol.*, 37, 357, 1962.
30. **Sanderson, C., Emmanuel, J., Emmanual, J., and Campbell, P.,** A historical review of paraffin and its development as an embedding medium, *J. Histotech.*, 11, 61, 1988.
31. **Scholer, J. and Armstrong, W. E.,** Aqueous aldehyde (Faglu) histofluorescence for catacholamines in 2 μm sections using polyethylene glycol embedding, *Brain Res. Bull.*, 9, 27, 1982.
32. **Smithson, K. G., Macvicar, B., and Hatton, G. I.,** Polyethylene glycol embedding: a technique compatible with immunocytochemistry, enzyme histochemistry, histofluorescence, and intracellular staining, *J. Neurosci. Methods,* 7, 27, 1983.
33. **Wolosewick, J. J.,** The application of polyethylene glycol (PEG) to electron microscopy, *J. Cell Biol.*, 86, 675, 1980.

Chapter 8

DIETHYLENE GLYCOL DISTEARATE (DGD): AN ALTERNATIVE TO PEG AS A REMOVABLE EMBEDMENT

David G. Capco

TABLE OF CONTENTS

0-8493-4323-2/93/$0.00 + $.50

I. INTRODUCTION

A useful alternative to polyethylene glycol (PEG) as a removable embedding medium is diethylene glycol distearate (DGD), which is wax-like and can be used for preparation of sections for observation by both light microscopy (LM) and electron microscopy (EM). Recent interest in this removable embedding medium stemmed from a report published in 1984 by Capco and Penman,[1] which demonstrated the utility of DGD for ultrastructural analysis of both the cytoskeleton and nuclear matrix. However, this report did not represent the first use of DGD for cytologic analysis. DGD was used for light microscopic analysis much earlier.[2-6] Unfortunately, the variable quality of commercially available DGD, with varying levels of impurity in different batches of the wax, affected the sectioning properties of different batches of wax, so that DGD was abandoned by many investigators.[7] Some investigators attempted to modify the sectioning properties of DGD with additives such as castor oil or cellulose caprate,[7,8] but these modifications were of limited value, possibly because of the variable purity of DGD from different suppliers. Polysciences, Inc. (Warrington, PA) has consistently supplied DGD with excellent sectioning characteristics since 1982, and this supplier served as the source of DGD for most of the recent investigations conducted with this removable embedding medium (see following section). The reliability of Polysciences in providing purified DGD is likely to have significantly contributed to the increasing number of investigations using DGD as a removable embedding medium. Recently, EM Corporation (Chestnut Hill, MA) has also begun supplying DGD as an embedding medium for ultrastructural analysis of biological samples.

Purified DGD is a hard, brittle, and waxy solid which can support sections of 40 nm thickness and can also be sectioned to a thickness greater than 1 μm. The additives, which some investigators used previously[7,8] alters the brittleness of the wax and adversely alters the sectioning characteristics of relatively pure DGD for thin sections; therefore, the author does not recommend these additives, particularly for ultrastructural studies. Addition of dimethylsulfoxide (DMSO; 0.3% final concentration) into DGD does not alter the sectioning characteristics of the wax, but it promotes penetration of the wax into biological material and reduces contraction of the embedment during the process of solidification.[9]

Some investigators use embedment-free sections to examine the detergent-resistant cytoskeleton of cells.[9-14] In this procedure living cells are exposed to nonionic detergent in a medium that stabilizes cytoskeletal components[15] prior to fixation. However, detergent extraction is not a prerequisite for use of this embedment, and unextracted cells have been successfully used with DGD embedment for light and electron microscopic analyses.[16-18] Without detergent extraction the cytoskeleton cannot be clearly visualized, though the cytomatrix of the cell is easily detectable.[16] In addition, in the absence of detergent extraction, the organelles of the cell remain intact and can be clearly visualized at the ultrastructural level. Whether cells are processed intact or after detergent extraction is dependent on the cellular component(s) of interest to the investigator.

II. METHODOLOGY FOR USE OF DGD

A. FIXATION

Any standard fixative for LM or EM can be used for embedding in DGD. The choice of fixative is dependent on the goals of the investigation. For example, if sections are to be used for immunocytochemical localization or *in situ* hybridization, then the choice of fixative, concentration of fixative, and length of fixation should be modified to allow the necessary chemical reaction to occur within the sections. Examples of fixatives for these purposes are described in other chapters in this book.

B. EMBEDDING IN DGD

After fixation, dehydration is performed through a graded ethanol series (the length of time spent in any one concentration of ethanol is dependent on the size of the specimen being dehydrated). A minimum incubation time of 2 h with several changes of 100% ethanol should be conducted before proceeding into the transition fluid. The transition fluid is 1-butanol, and a graded series should be used to transfer the specimen from 100% ethanol (at least a 1:1 mixture of 1-butanol to 100% ethanol). Subsequently, the specimen is transferred into 100% 1-butanol with several changes of 100% 1-butanol over a minimum of 2 to 3 h. 1-Butanol slowly dissolves some plastics, thus processing should be done either in Nalgene® plastic, in glass tubes, or glass Petri plates. During the final change of 1-butanol the sample should be placed in an oven adjusted to 70°C for about 15 min to warm the tube and the 1-butanol prior to infiltration into the embedding medium.

Solid DGD should be melted in an oven at 70°C prior to use. It is recommended that the DGD be made 0.3% with DMSO prior to use, as this reduces the amount of contraction which occurs in solidification of the block. After adding DMSO, the mixture should be thoroughly agitated, and in this state it can be resolidified and remelted multiple times. Occasionally, a white flocculent will appear in the molten DGD, particularly if it has been left in the molten state for several days. This can be removed by filtration through a Whatman #1 filter conducted in an oven.

A 1:1 premixed solution of 1-butanol:molten DGD is added to the samples and allowed to incubate for at least 45 min. Subsequently, this 1:1 mixture is replaced with several changes of DGD (containing 0.3% DMSO) and allowed to incubate for a minimum of 2 h (dependent upon the size of the specimen). At this stage, the specimen can be subjected to vacuum infiltration in a vacuum oven to promote more thorough penetration of DGD into the tissue if it is needed.

Infiltration in flat embedding molds is recommended. This is because once solidified, DGD becomes very brittle and the pressure needed to remove the solidified DGD block containing the specimen from an embedding mold, such as BEEM™ capsule, can cause enough strain on the block that it fractures and subsequently crumbles during block trimming or sectioning. Flat embedding molds with a depth of 4 mm should be used, as this provides enough DGD to compensate for any contraction that might occur during solidification, and the block remains shallow enough so that it can be removed from the embedding mold without causing fracture of the block. If it is necessary, orient the specimen within the embedding mold prior to the beginning of the solidification process. This can be done by placing heat lamps around a dissecting microscope to heat the stage while orienting the specimen. The mold is then removed from the area containing the heat lamps to cool. As the block cools it will begin to contract slightly, and a little molten DGD should be added to the edge of the mold opposite to the specimen during the process of solidification, so that when the block is completely solidified it has a fairly flat upper surface. Do not place the embedding mold on ice or in the refrigerator to promote rapid solidification, as this process usually causes such rapid temperature changes that the block fractures. Instead, the block should be left to cool at room temperature. To remove the block from the embedding mold after it has solidified (a process usually requiring about 20 min) the edges of the mold surrounding the block should be flexed slightly to cause the adhesion points between the block and the mold to break. After the block has been thoroughly loosened along its sides, it should be lifted out by bending the mold. Avoid causing any stresses on the block, otherwise it will crack. Note that once the DGD has solidified, remelting will damage the specimen for ultrastructural analysis.

C. PREPARATION OF DGD SECTIONS

The block is placed in a vice-type chuck for microtomy. Use a razor blade to trim the upper surface of the block to flatten it so that it fits securely in the vice grip. Trimming can be performed as with resin-embedded blocks, except it is necessary to make shallow cuts in trimming the block, since a deep cut can cause enough stress to fracture the block. It is preferable to start with specimens that are small enough that a layer of DGD can be left around the specimen. The DGD surrounding the specimen during sectioning also helps promote the formation of ribbons and appears to enhance the integrity of the specimen as it floats on the water. The hardness of block is dependent on the temperature of the room. It is preferable to section in a room that is approximately 22°C or cooler. If the block warms because of the room temperature or because of heating from the light source of the microtome, it is essential to pause to allow the block to cool and reestablish the proper conditions for obtaining good sections. Do not touch the block with fingers in the region of the specimen, as this softens the block. Once trimmed, sections can be obtained as with any standard resin block, except that the knife angle should be set at 10°. Sections floated on water-filled troughs will provide interference colors to give an indication of the section thickness and the relative thickness of one section to another. The sequence of interference runs from silver, to gold, to green, to blue, then to dark gold, to dark green, and to dark blue. A similar sequence will repeat itself for an additional cycle. The integrity of silver sections is rarely adequate; however, it is possible to obtain acceptable gold sections. Because the embedding medium will be removed, it is also possible to use sections much thicker than this. For example, the subsequent blue or green or even the dark gold sections can provide excellent images of the tissue of interest at the ultrastructural level. In addition, sections with a thickness greater than 1 μm can be prepared for LM observation. It is better to prepare sections of increasing thickness until you determine the maximal thickness that can be used. For example, in detergent-extracted specimens for ultrastructural analysis[1] sections over 0.5 μm thick can still be easily penetrated by an electron beam with an accelerating voltage of 80 kV.

The sections float and form ribbons on the water surface of trough. These ribbons can be picked up on grids that are formvar-coated, carbon-stabilized, and treated with 0.1% polylysine (380 kDa mol wt). Polylysine should be applied to the grid immediately before use. A problem often encountered is that during the removal of the embedding medium from the sections, some of the sections float off the grid, even though the grids are polylysine coated. To improve adhesion of the sections to the grids it is recommended that after their initial collection they have to be dried completely and then a droplet of polylysine is added to the grid over the sections and allowed to incubate for several additional minutes. The excess polylysine is then removed and the specimen dried again overnight in a vacuum jar containing desiccant.

D. REMOVAL OF EMBEDDING MEDIUM FROM SECTIONS

After thoroughly drying the specimens on the supporting carriers (i.e., grids, slides, or coverslips), the DGD is removed from the sections by immersing the loaded carriers in 1-butanol. This should be done as gently as possible, so as not to dislodge the sections from the carrier while it is being immersed. For sections with a thickness of less than 400 nm, complete removal of DGD takes about 1 h. Thicker sections require longer incubation times in the 1-butanol. The carriers should be agitated one time by gently swirling the solution of 1-butanol and then allowed to incubate an additional 15 min prior to transferring the support to the next solution. The next solution is a mixture of 100% 1-butanol and 100% ethanol. A minimum incubation of 15 min in that solution is required. Subsequently, the support is transferred to a solution of 100% ethanol and allowed to sit for a minimum of 1 h, swirled gently, and allowed to sit for an additional 15 min.

For light microscopic observations, the specimens on slides or coverslips can be histochemically stained and mounted. Alternatively, sections can be processed for analysis by any method applicable to paraffin-embedded thick sections, such as immunocytochemistry (ICC) or *in situ* hybridization (see following). For ultrastructural analysis, specimens on grids can be placed in a holder for critical point drying, subsequently dried through the CO_2 critical point, and examined in the transmission electron microscope (TEM) as an embedment-free section. Grids should be retained in a vacuum desiccator until use. During observation it is recommended that the emission current of the microscope be reduced to half of what would be typically used for a resin-embedded section and that the cold trap be completely filled. This appears to promote stability of the specimen in the beam. In case the specimen remains unstable, a thin layer of carbon can be evaporated onto the surface of the specimen prior to observation.

Embedment-free sections have particularly high contrast and small structures such as filaments present on negatives are not transfered to the print without overexposure. We found it necessary to adjust the developing conditions to provide an image which is printable. For this purpose, Kodak® EM film emulsion number 4489 should be developed in a freshly made solution of Kodak® D-19, diluted 1:8 with water, and developed for a total of 8 min. This developer is used only once (i.e., discard after development of one batch of negatives). Washing and fixation follow standard procedures. If the contrast of the negatives remains too great, it may be necessary to further modify the development time or the exposure time for the negative.

Post-staining with heavy metal stains for ultrastructural analysis is not necessary; these specimens have intrinsically high contrast because there is no embedding resin present, and the image is created by the scattering of electrons by the biological material in a vacuum. For similar reasons post-fixation with osmium tetroxide is not needed to provide contrast and can be omitted if desired. However, fixation with osmium promotes specimen stability under the electron beam and for this reason it is recommended.

III. APPLICATION OF DGD EMBEDDING METHOD FOR CYTOLOGIC STUDIES

A. USE IN TRANSMISSION ELECTRON MICROSCOPY STUDIES

One of the useful features of DGD as an embedment is that sections of this material float on water-filled troughs of glass or diamond knives, whereas PEG sections must be cut with a dry knife. Moreover, interference colors are produced as light reflects between the DGD section and the water-filled trough when the trough is illuminated from above. Thus, as with plastic-embedded sections, it is possible to estimate section thickness based upon the interference color that results and to select for sections of consistent thickness using these colors. Another useful feature of DGD as an embedment, which also is an advantage for PEG, is that sections significantly thicker than plastic-embedded sections can be analyzed in the electron microscope. Once the embedding medium is removed, the biological material can be penetrated with an electron beam generated by an accelerating voltage of 60 to 100 kV of any conventional transmission electron microscope. The thickness limitation for plastic-embedded sections results from the fact that the electron beam is deflected by the embedding plastic. In the absence of any embedding medium the electron beam can easily penetrate biological material, and a thickness limitation results only when the superimposition of three-dimensional biological material becomes too confusing to interpret. Microscopists, whose goal in thin sectioning has always been to obtain the thinnest section possible, can redefine that goal when preparing embedment-free sections. Thick, embedment-free sections allow acquisition of more spatial cues for three-dimensional reconstruction of the extracellular matrix and the cell interior, which also provide clear views of the organization of cytoskeletal components. A very thin, embedment-free section cannot provide the same amount of spatial information, and there is little advantage to the microscopist using such a section.

Thin sections cut through virtually all of the organelles of the cell. In contrast, thicker sections, which can be analyzed after removal of the embedding medium, frequently capture entire cell organelles within the plane of the section. Consequently, cell components in thick, embedment-free sections appear different than the familiar textbook-documented images of mitochondria, vesicles, Golgi apparatus, and other cell organelles, because entire organelles are observed from the exterior rather than as a thin two-dimensional plane through the interior of a three-dimensional object. There are, however, limitations to the use of embedment-free sections in the TEM. One limitation, already mentioned above, is that as thickness increases, the superimposition of an increasing number of cellular components can result in an uninterpretable image. A second limitation of the method is that at high magnification the resolution of the image decreases. In embedment-free sections the image is formed by deflection of electrons in the electron beam by the biological material, whereas no deflection occurs in the vacuum surrounding the specimen. When examining a minute feature of a cellular component at high magnification the image quality can decrease because there is insufficient electron deflection to resolve the component. Under these circumstances a different method of microscopy would be recommended, such as negative staining or rotary shadowing (see Chapter 4).

Specimens can be processed as embedment-free sections without detergent extraction, or they may be processed after detergent extraction and analyzed using TEM.[1,16] Figure 1A is a low-magnification view of salivary gland cells from *Drosophila* showing a nucleus containing polytene chromosomes. This preparation was not detergent extracted, and consequently the numerous secretory granules of these excretory cells can be clearly visualized. Figure 1B is a higher-magnification view of polytene chromosomes within the nucleus. Note that the banding pattern of the chromosomes is clearly visible and does not require heavy metal staining. Typically the banding pattern of these chromosomes is visualized after making a ''squash'' preparation and subsequently staining the preparation. Such preparations are typically of low resolution and destroy the three-dimensional structure of both the nucleus and cytoplasm. Embedment-free sections allow high resolution examination of

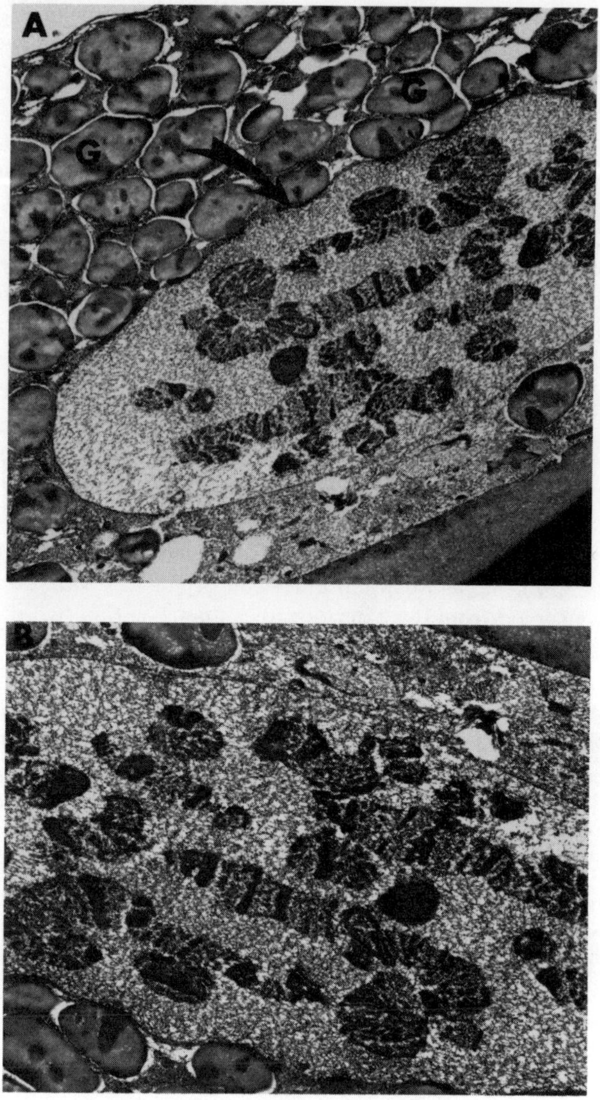

FIGURE 1. Embedment-free section of salivary gland cell from *Drosophila* salivary gland. Intact, nondetergent extracted salivary gland was embedded in DGD, sectioned at 0.25 μm thickness and processed for observation in TEM as described in text. (A) Low magnification view shows polytene chromosomes within nucleus (arrow). Cytoplasmic area is filled with secretory granules (G) of this secretory cell. Note that banding pattern of the chromosomes can be detected without the aid of DNA stain. (Magnification × 3,400.) (B) Higher magnification (× 6,750) view demonstrates that chromosomes are enmeshed in protein-RNA matrix of nucleus. (From Capco, D. G., unpublished observations.)

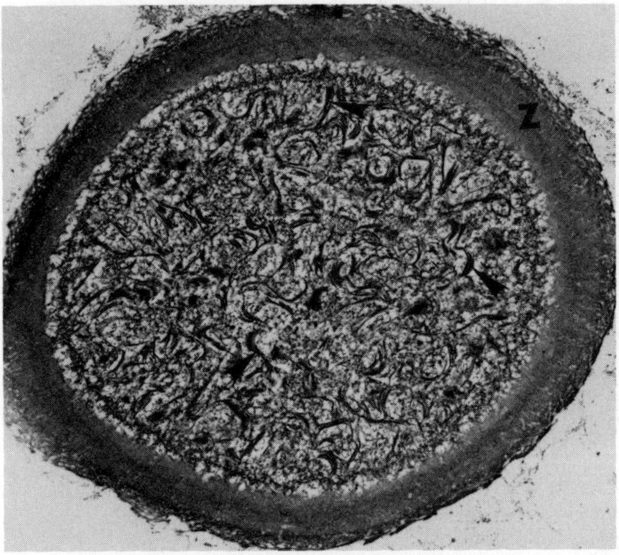

FIGURE 2. Low magnification, embedment-free section of detergent-extracted hamster egg. Cytoskeletal sheets (arrowheads) unique to mammalian embryos can be seen in egg interior. Thin filament network typically found in somatic cells surround sheets and is enriched in egg cortex. Egg is surrounded by specialized extracellular matrix, zona pellucida (Z). (Magnification × 1,800.) (From McGaughey, R. W. and Capco, D. G., unpublished observations.)

interactions between the chromosomes and surrounding nuclear components and also allow detection of cytoplasmic specializations in the cytomatrix that associate with the nuclear perimeter.

Ultrastructural analysis of embedment-free sections from mammalian eggs that have been detergent extracted revealed unique cytoskeletal elements within these specialized cells.[9,12] These cytoskeletal elements, seen in Figure 2, are referred to as cytoskeletal sheets, and appear to play an important role in both the development of embryonic polarity and in the formation of the cellular precursor to the placenta.[9,13,14] The functional role of these sheets was completely unanticipated until thick, embedment-free sections were examined ultrastructurally. The thick sections provided spatial cues that allowed interpretation of these unique cellular components,[14,19,20] which otherwise would have been unrecognized.

Embedment-free sections obtained from DGD embedding processes have played a central role in dissecting the structure and function of the nuclear matrix and the nuclear core filaments. Work on these structural components has bene recently reviewed by Nickerson and Penman (1991).[21] These structural elements of the nucleus serve as attachment points for the DNA, give shape to the nucleus, and reorganize as the cell enters mitosis.[10,21-24] The nuclear core filaments shown in Figure 3 have been described by Nickerson

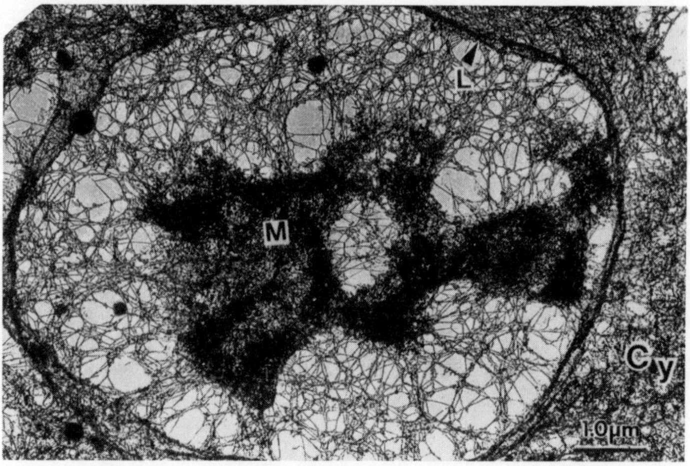

FIGURE 3. Embedment-free section of nuclear core filaments. Nuclear core filaments provide armature upon which rest of nuclear matrix is organized. Cy, cytoskeleton; L, nuclear lamin; M, dark mass which is remnant of nucleolus. (Magnification × 10,000.) (From He, D., Nickerson, J. A., and Penman, S., *J. Cell Biol.*, 110, 569, 1990. With permission.)

and Penman[21] as the armature around which the other components of the nuclear matrix are assembled. The protein composition and structure of the nuclear matrix are highly specific and can serve to distinguish different cell types.[10] Moreover, the protein composition of the nuclear matrix is rapidly altered in response to heat shock[25,26] and in response to viral infection,[27,28] suggesting that the nuclear matrix is involved in a rapid metabolic response to cellular traumas. Figure 4A illustrates the nuclear matrix in CHO cells prior to heat shock, and Figure 4B shows the structural change after heat shock (also see Ornelles and Penman, 1990[26]). The structural changes in the nuclear matrix would be impossible to detect in the studies described previously without the use of embedment-free sections, as the superimposition of the cytoskeleton enveloping the nucleus, and also the nuclear lamina at the periphery of the nucleus, would obliterate structural changes in the nuclear interior.

B. USE IN IMMUNOLOCALIZATION STUDIES

Embedment-free sections made from specimens embedded in DGD have been used for immunofluorescent analysis by Lacy and Davies (1959),[4] Valdimarsson and Huebner (1989),[18,29] and Ojeda et al. (1986),[30] and ultrastructural immunodetection by Nickerson et al. (1990)[23] and Capco et al.[31] For immunofluorescent analysis the sections were prepared as described previously for ultrastructural analysis, and 1 to 2 μm-thick sections were placed on polylysine-coated coverslips or slides. The embedding medium was removed by a 3 h incubation in 1-butanol (thicker sections require longer

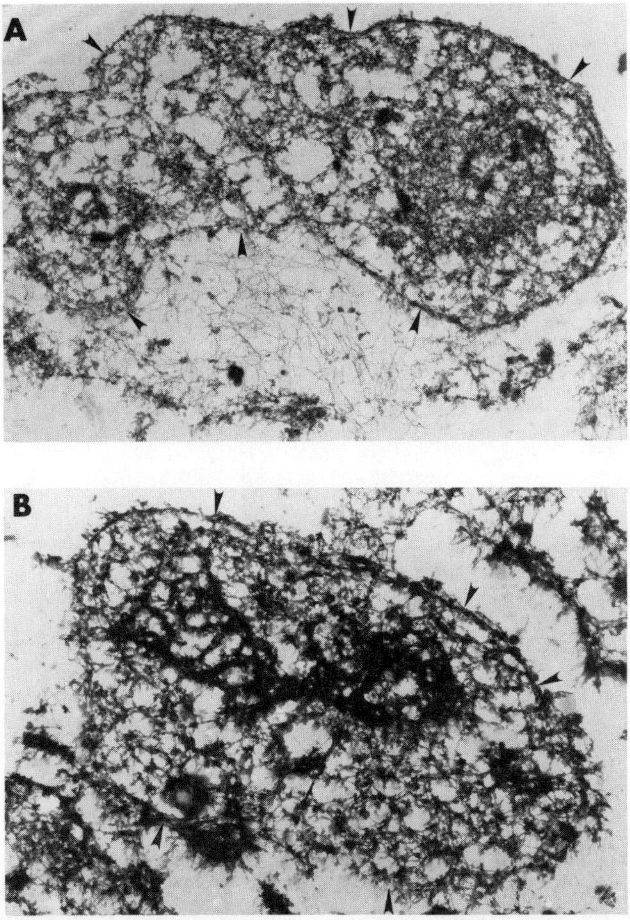

FIGURE 4. Embedment-free section of nuclear matrix in CHO cells before and after heat shock. Nuclear periphery is indicated with arrowheads. (A) Prior to heat shock nuclear matrix contains numerous fine filaments. (B) After heat shock structure of nuclear matrix is altered. Instead of fine filaments, thick fibers are present which anastomose with neighboring filaments. (Magnification × 9,000.) (From Phyllis Wachsberger and Ronald Cross, Radiation Oncology and Nuclear Medicine, Thomas Jefferson University, unpublished observations.)

incubation times to remove the DGD) by one group of investigators,[18] while another group removed DGD by incubation in xylene.[30] Once the DGD was removed, specimens were transferred to 100% ethanol and rehydrated through a graded ethanol series of decreasing ethanol concentrations. Standard procedures for blocking, washing, challenge with primary and secondary antibody, and observation were employed by these investigators.[18,29,30] Figure 5 demonstrates the immunofluorescent localization of microtubules in a section of an insect ovary. Ojeda et al. (1989)[30] employed sections at varying thickness

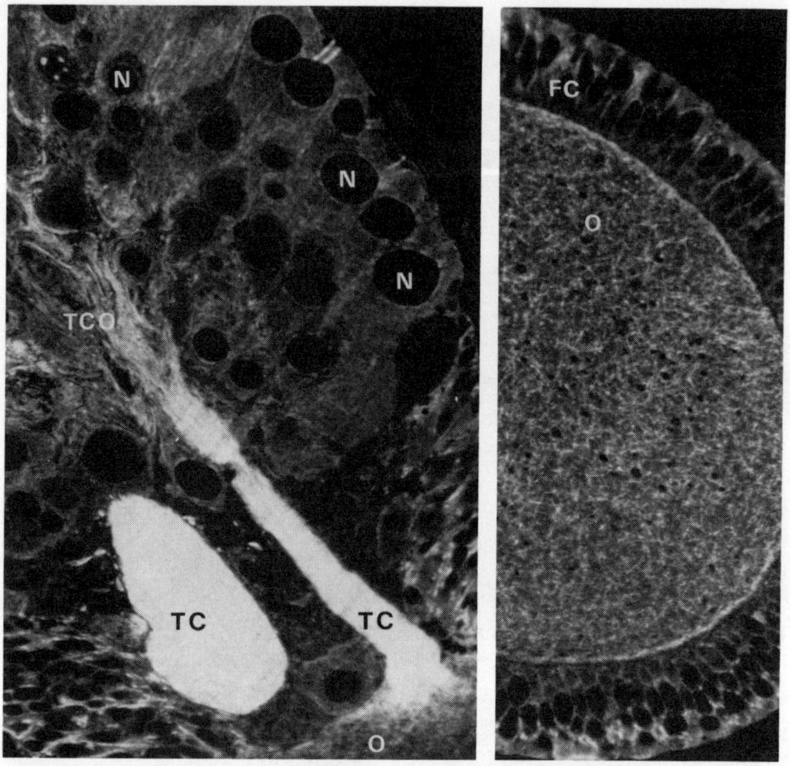

FIGURE 5. Immunofluorescent localization of tubulin in embedment-free section of meriostic ovary of an insect. (A) Tubulin is localized in trophic cords (TC) and trophic core (TCO) of syncytium composed of nurse cells (N). Connection of trophic core to oocyte (O) exists in lower right corner of frame. (B) View of oocyte (O) surrounded by follicle cells (FC). Microtubules form space-filling network within the oocyte. (Magnification × 600.) (From Valdimarsson G. and Huebner, E., *Biochem. Cell Biol.*, 67, 242, 1989a. With permission.)

down to 0.5 μm thick. They observed that when thinner sections were used for immunofluorescent analysis, less image deterioration occurred due to light scatter from out-of-focus parts of the specimen. They noted that while computerized video image processing or confocal laser microscopy can subtract the light scatter from the out-of-focus portion of the image, it is an expensive technology to apply, but thin embedment-free sections prepared from DGD can produce the same results.

Immunodetection at the ultrastructural level is also possible in embedment-free sections prepared from DGD. For this process, the sections are collected on gold or nickel grids coated with formvar film and stabilized with carbon. The DGD is removed with 1-butanol and transferred into 100% ethanol as described above for ultrastructural with 1-butanol and transferred into 100% ethanol as described above for ultrastructural analysis. However, instead of critical point drying, the specimens on the grids are rehydrated in a graded ethanol series of decreasing ethanol concentration and transferred into phosphate-buffered saline (PBS). At this point any standard procedure for immunodetection at the ultrastructural level can be employed. At the end of the final washing regime, the grids are incubated in 2% glutaraldehyde to immobilize the gold conjugated Protein A (or second antibody), washed again, and the specimen is dehydrated through a graded ethanol series and dried through the CO_2 critical point.[23,31]

C. USE IN *IN SITU* HYBRIDIZATION STUDIES

Sections prepared from tissue embedded in DGD can be hybridized *in situ* with nucleic acid probes using techniques developed for paraffin-embedded specimens.[17] Only the method of preparing sections differs from conventional wax-embedded specimens. Because sections of DGD can be prepared thinner than many wax-embedded specimens, the researcher has the option to increase the resolution of the images processed for *in situ* hybridization by employing sections with a thickness of 1 μm or less. DGD is removed from sections which are adhered to polylysine-coated slides as described in the methods section and transferred into 100% ethanol. Subsequently, sections are rehydrated by passage through a graded ethanol series of decreasing ethanol concentration. These sections can then be processed using any standard methodology for *in situ* hybridization.

IV. CONCLUDING REMARKS

Diethylene glycol distearate (DGD) as an embedment has a broad spectrum of applications in cytologic studies for both light and electron microscopic analyses. Problems with the variable quality of commercially available DGD that plagued studies prior to 1980 appear to have been solved, and at least one supplier, Polysciences Inc., has supplied DGD of excellent quality for the past decade. For studies involving light microscopy, sections prepared using DGD as the embedment can be used for any application that paraffin sections are typically used. For ultrastructural studies DGD sections can be prepared as easily as plastic-embedded sections, and use of embedment-free sections allows analysis of sections much thicker than plastic-embedded material. DGD will not infiltrate hard components, such as yolk platelets of amphibian embryos (Capco, unpublished results). In this regard we have found that PEG 400 distearate,[32] not to be confused with PEG 400, modified from Steedman's Polyester Wax,[33] performs an excellent job of infiltrating yolk platelets of amphibian eggs. We have used this embedment to prepare sections for LM analysis.[34]

Recently, some investigators have begun to refer to DGD as "Pentament" in recognition of Professor Sheldon Penman's rediscovery of this embedment early in the 1980s[1] and also in recognition of his numerous insightful contributions to the investigation of cellular architecture[35-39] and organization of the nuclear matrix.[1,10,11,21,23,25,26,28,38,40] This author wholeheartedly supports the idea of referring to this embedment as Pentament.

REFERENCES

1. **Capco, D. G., Krochmalnic, G., and Penman, S. A.,** A new method of preparing embedment-free sections for TEM: applications to the cytoskeletal framework and other three-dimensional networks, *J. Cell Biol.*, 98, 1878, 1984.
2. **Cutler, O. I.,** Embedding in glycol stearate, *Arch. Pathol.*, 20, 445, 1935.
3. **Chesterman, W. and Leach, E. H.,** A modified ester wax for embedding tissues, *O. J. Microsc. Sci.*, 97, 593, 1956.
4. **Lacy, P. E. and Davies, J.,** Demonstration of insulin in mammalian pancreas by the fluorescent antibody method, *Stain Technol.*, 34, 85, 1959.
5. **Salazar, H.,** Diethylene glycol distearate embedding and ultramicrotome sectioning for light microscopy, *Stain Technol.*, 39, 13, 1964.
6. **Taleporos, P.,** Diethylene glycol distearate as an embedding medium for high resolution light microscopy, *J. Histochem. Cytochem.*, 22, 29, 1974.
7. **Taleporos, P.,** The composition and cutting properties of diethylene glycol distearate from various sources, *J. Histochem. Cytochem.*, 24, 1285, 1976.
8. **Graham, E. T.,** Improved diethylene glycol distearate embedding wax, *Stain Technol.*, 57, 39, 1982.
9. **Capco, D. G. and McGaughey, R. W.,** Cytoskeletal reorganization during early mammalian development: analysis using embedment-free sections, *Develop. Biol.*, 115, 446, 1986.
10. **Fey, E. G., Krochmalnic, G., and Penman, S.,** The monchromatin substructures of the nucleus: The ribonucleoprotein (RNP)-containing and RNP-depleted matrices analyzed by sequential fractionation and resinless-section electron microscopy, *J. Cell Biol.*, 102, 1654, 1986.
11. **He, D., Nickerson, J. A., and Penman, S.,** The core filaments of the nuclear matrix, *J. Cell Biol.*, 110, 569, 1990.
12. **McGaughey, R. W. and Capco, D. G.,** Specialized cytoskeletal elements in mammalian eggs: structural and biochemical evidence for their composition, *Cell Motil. Cytoskel.*, 13, 104, 1989.
13. **Gallicano, G. I., McGaughey, R. W., and Capco, D. G.,** The cytoskeleton of the mouse egg and embryo: reorganization of planar elements, *Cell Motil. Cytoskel.*, 18, 143, 1991.
14. **Gallicano, G. I., McGaughey, R. W., and Capco, D. G.,** Cytoskeletal sheets appear as universal components of mammalian eggs, *J. Exp. Zool.*, 263, 194.
15. **Penman, S., Capco, D. G., Fey, E. G., Chatterjee, P., Reiter, T., Ermisch, S., and Wan, K.,** The three dimensional structural networks of cytoplasm and nucleus; Function in cells and tissue, in *The Modern Cell Biology Series: Spatial Organization of Eukaryotic Cells*, McIntosh, J. R., Ed., Alan R. Liss, New York, 385, 1983.
16. **Walch, E. T., Martin, V. J., and Archer, W. E.,** Evidence of a microtrabecular cytoskeletal lattice in glandular cells of hydrozoan Planulae, *J. Morphol.*, 187, 353, 1986.
17. **Korenbrot, J. I. and Fernald, R. D.,** Circadian rhythm and light regulate opsin mRNA in rod photoreceptors, *Nature*, 337, 454, 1989.
18. **Valdimarsson, G. and Huebner, E.,** Diethylene glycol distearate as an embedding medium for immunofluorescence microscopy, *Biochem. Cell Biol.*, 67, 242, 1989a.
19. **Capco, D. G. and Larabell, C. A.,** The cytoskeleton during early development: structural transformation and reorganization of RNA and protein, in *Progress in Molecular and Subcellular Biology*, Jeanteur, P., Kuchino, Y., Muller, W. E. G., Paine, P. L., Eds., Springer-Verlag, Frankfurt, 59, 1991.
20. **Bement, W. M., Gallicano, G. I., and Capco, D. G.,** The role of the cytoskeleton in early development, *Microsc. Res. Tech.*, 22, 23, 1992.
21. **Nickerson, J. A. and Penman, S.,** BioVision: microscopy in three dimensions, *Sem. Cell Biol.*, 2, 117, 1991.

22. **Wagner, B., Krochmalnic, G., and Penman, S.,** Resinless-section electron microscopy of HeLa cell mitotic architecture, *Proc. Natl. Acad. Sci. U.S.A.*, 83, 8996, 1986.
23. **Nickerson, J. A., Krockmalnic, G., He, D., and Penman, S.,** Immunolocalization in three dimensions: immunogold staining of cytoskeletal and nuclear matrix proteins in resinless electron microscopy sections, *Proc. Natl. Acad. Sci. U.S.A.*, 87, 2259, 1990.
24. **Jiao, R., Wu, D., Zhang, B., Cai, S., and Zhai, Z.,** Immunogold labelling of the intermediate filament-lamina-nuclear matrix system in HeLa and BHK-21 cells, *J. Electron Microsc. Tech.*, 18, 126, 1991.
25. **Reiter, T. and Penman, S.,** "Prompt" heat-shock proteins: translationally regulated synthesis of new proteins associated with the nuclear matrix-intermediate filaments as an early response to heat shock, *Proc. Natl. Acad. Sci. U.S.A.*, 80, 4737, 1983.
26. **Ornelles, D. A. and Penman, S.,** Prompt heat-shock and heat-shifted proteins associated with the nuclear matrix-intermediate filament scaffold in Drosophila melanogaster cells, *J. Cell Sci.*, 95, 393, 1990.
27. **Weed, H. G., Krochmalnic, G., and Penman, S.,** Polio virus metabolism and the cytoskeletal framework: detergent extraction and resinless section electron microscopy, *J. Virol.*, 56, 549, 1985.
28. **Zhonghe, Z., Nickerson, J. A., Krochmalnic, G., and Penman, S.,** Alterations in nuclear matrix structure after adenovirus infection, *J. Virol.*, 61, 1007, 1987.
29. **Valdimarsson, G. and Huebner, E.,** The development of microtubular arrays in the germ tissue of an insect telotrophic ovary, *Tissue and Cell,* 21, 123, 1989b.
30. **Ojeda, J. L., Ros, M. A., and Icardo, J. M.,** A technique for fluorescence microscopy in semithin sections, *Stain Technol.*, 64, 243, 1989.
31. **Capco, D. G., Gallicano, G. I., McGaughey, R. W., Downing, K. H., and Larabell, C. A.,** Cytoskeletal sheets of mammalian eggs and embryos: a lattice-like network of intermediate filaments, *Cell Motil. Cytosk.*, 24, in press.
32. **Norenburg, J. L. and Barrett, J. M.,** Steedman's polyester wax embedment and deembedment for combined light and scanning electron microscopy, *J. Electron Microsc. Tech.*, 6, 35, 1987.
33. **Steedman, H. F.,** A new ribboning embedding medium for histology, *Nature,* 179, 1345, 1957.
34. **Bement, W. M. and Capco, D. G.,** Parallel pathways of cell cycle control during Xenopus egg activation, *Proc. Natl. Acad. Sci.*, 88, 5172, 1991.
35. **Lenk, R., Ransom, L., Kaufmann, Y., and Penman, S.,** A cytoskeletal structure with associated polyribosomes obtained for HeLa cells, *Cell,* 10, 67, 1977.
36. **Lenk, R. and Penman, S.,** The cytoskeletal framework and poliovirus metabolism, *Cell,* 16, 289, 1979.
37. **Fey, E. G., Wan, K. W., and Penman, S.,** Epithelial cytoskeletal framework and nuclear matrix-intermediate filament scaffold: three dimensional organization and protein composition, *J. Cell Biol.*, 98, 1973, 1984.
38. **Penman, S.,** Virus metabolism and cellular architecture, in *Virology,* Fields, B. N., Ed., Raven Press, New York, 169, 1985.
39. **Penman, S.,** If genes just make proteins and our proteins are the same, then why are we so different? *J. Cell Biochem.*, 47, 95, 1991.
40. **Capco, D. G., Wan, K. M., and Penman, S.,** The nuclear matrix: three-dimensional architecture and protein composition, *Cell,* 29, 847, 1982.

INDEX

E

F

G

M

N

O

P

R

S

T

U

V

W

X